AF505959

THE REFORM OF TIME

The Reform of Time

Magic and Modernity

Maureen Perkins

Pluto Press

LONDON • STERLING, VIRGINIA

First published 2001 by Pluto Press
345 Archway Road, London N6 5AA
and 22883 Quicksilver Drive,
Sterling, VA 20166–2012, USA

www.plutobooks.com

British Library Cataloguing in Publication Data
A catalogue record for this book is available from
the British Library

Library of Congress Cataloging in Publication Data
Perkins, Maureen.
    The reform of time : magic and modernity / Maureen Perkins
        p. cm.
    ISBN 0–7453–1729–4 (hbk.) — ISBN 0–7453–1728–6 (pbk.)
    1. Time—Sociological aspects. 2. Time management—History—19th
century. 3. Time measurements—History—19th century. 4.
Superstition—History—19th century. 5. Magic—History—19th century.
6. Prophecy—History—19th century. I. Title.
    HM656 .P47 2001
    304.2'3—dc21
                                                            00–011095

10   09   08   07   06   05   04   03   02   01
10   9   8   7   6   5   4   3   2   1

ISBN 0 7453 1729 4 hardback
ISBN 0 7453 1728 6 paperback

Designed and produced for Pluto Press by
Chase Publishing Services
Typeset from disk by Stanford DTP Services, Northampton
Printed in the European Union by TJ International, Padstow

Contents

List of Figures

Acknowledgements

Many readers, so I'm told, turn to the Acknowledgements on first opening a book. I share this interest in someone else's friendships and debts, though perhaps with a faint trace of guilt at reading the least 'important' bit of a book first, and of allowing it to influence my expectations about what comes next. Is the popularity of this practice because readers hope to find the author's true voice - to give a special insight into what follows? If that is so, then I hope that the reader of *The Reform of Time* will find evidence of something 'personal' in more than just these few lines. At several points in the body of the text I have tried to convey a sense of debt to modern postcolonial theory, which has in recent years offered me new and empowering ways of formulating a lifetime's experiences of race, gender and class. At the same time I have tried to practise an empirically-based historical methodology, respecting the specificity of particular contexts. Debates within the academy have sometimes suggested that these two endeavours – theory and empiricism – might be mutually antagonistic. If I have managed at all to combine an interest in cultural theory with a regard for accurate historical facts founded on textually based research, it will be due in large part to Roger van Zwanenberg, who has encouraged me from the outset to be more courageous in my claims. I am enormously grateful to him for hearing an implicit politics in my earlier work and urging me to make it more explicit. If I still fail to be quite as brave as he would like, the work is in progress.

My friend and colleague Rob Stuart defends the footnote as a signal that scholarly work is never a solitary venture, but rather builds on the achievements of others. In this regard, there would be too many people to name here, but I would like to single out especially Patrick Curry, Iain McCalman, F.B. Smith, Peter Read and Michael Roberts. All have given me support and encouragement, and all have written histories that I have found inspiring, which have helped me to ask new questions and to make new connections.

During the writing of this book, I was the fortunate recipient of first, a University of Western Australia postdoctoral fellowship, and then an Australian Research Council postdoctoral fellowship. During that time, I was greatly sustained by the warmth and generosity of Richard Bosworth and Frank Broeze. The staff of the Scholars' Centre of the University of Western Australia were simply wonderful; especial thanks to Susana Melo de Howard and Toby Burrows; and to Trudi McGlade, for cheerful, generous guidance about administration. The Scholars' Centre allowed me to reproduce Figure 11 from a copy of *Punch* held by the Humanities and Social Sciences Library, University of Western Australia. Other *Punch* illustrations were reproduced with the permission of the Battye Library, Western Australia (Figures 12, 13, 14, 18, and 19). Figures 1, 5–10, and 15 are reproduced by permission of the British Library. The National Maritime Museum, Greenwich, gave permission for the reproduction of Figure 2; Cambridge University Library for Figure 3; the Museum of Contemporary Art, Los Angeles, for Figure 4; and the Louvre, Paris, for Figure 16. My thanks to all these institutions and their kind staff.

The research was also helped by a visiting fellowship at the Centre for the Book at the British Library, as well as two grants from the Australian Research Council. Thanks also to Robert Poole, Linda Persson, the Social History Society, and Owen Davies, all for making Perth a little less isolated; and to Robert Webb, for helpful support at Pluto Press, again at a distance.

To all those who have read or commented on versions of these chapters many thanks, especially Rob Stuart, Trish Crawford, Robert Poole, Victoria Burbank, Juliette Peers, Judy Berman, and Anna Davin. Earlier versions of chapters 3 and 4 appeared in journals, and I would like to thank their anonymous readers. Parts of chapter 3 first appeared in 'The Meaning of Dream Books' in *History Workshop Journal*, 48 (Autumn 1999), pp. 102–13. Chapter 4 includes a version of my article 'Timeless Cultures: the "Dreamtime" as colonial discourse,' reprinted by permission of Sage Publications Ltd, from *Time & Society*, 7 (2), 1998, pp. 335–51.

Thank you to Cheryl Lange, Jean-Marie Volet and Tony and Jo Barnes for support and encouragement. And thank you, as always, to Rob, Matthew, and Roberta.

Lastly, with feeling: any errors that remain are my own. I am mindful of the privilege of being able to risk making them.

Progress leaves its dead by the way, for progress is only a great adventure as its leaders and chiefs know very well in their hearts. It is a march into an undiscovered country, and in such an enterprise the victims do not count. (Joseph Conrad, 'The Crime of Partition', in *Notes on Life and Letters*, p. 18, quoted by Eloise Knapp Hay, *The Political Novels of Joseph Conrad* (Chicago: University of Chicago Press, 1963), pp. 174–5).

Introduction: Superstition and Progress

In the 1840s Samuel Bamford, who had been at various times a sailor, a weaver and a political radical, published his autobiography, one of several working-class memoirs to appear in Britain during the late eighteenth and early nineteenth centuries. Bamford's story is marvellously entertaining, the account of a young man rushing passionately through life: time spent in prison, escaping from the press-gang, his baby girl put into his arms the morning after his wedding day. Among the most distinctive features of the narrative is his interest in popular belief, or superstition. He reports a horseshoe put up over a door to prevent the entry of witchcraft, the disappearance of a man thought to be stolen by witches, the death of a young lad who dabbled in magic in order to make a girl fall in love with him. In all these stories, Bamford's own attitude is complex. He neither dismisses such accounts from an educated point of view, nor offers them unproblematically as authentic. A modern editor refers to his 'half-belief', and Bamford himself describes his role as that of 'an enquirer and listener' rather than one who disputed what he was told.[1] His ambivalent fascination with the supernatural is a perfect example of modern, rationalist ideas overlaying an earlier world of popular belief. In particular, the uneasy coexistence of fate and free will consistently appears as the central motif of these stories. Could the church sexton who saw his own ghost on All Souls' Eve (when the spirits of all who are to die in the coming year are said to appear) have escaped his own death? Was it destiny or his own uneducated fear which caused the unfortunate man to go home and fall ill? Was the awful fate of the young men who attempted magic in Boggart Hole Clough a sign to all that spirit meddling was wrong, or did their success in making the young woman fall in love testify to the power of their spells?[2]

Bamford's memoir highlights one of the persistent problems in conceptualising human agency, that of the interplay between determinism and free will. These Herculean pillars of philosophy define the parameters of discussions ranging across the whole

THE TRUE

FORTUNE TELLER;

OR

UNIVERSAL BOOK OF FATE.

Containing besides other valuable information, directions by which any one may know under what planet he was born.—An account of the evil and perilous days of every month of the year.—How to choose a husband or wife by the hair, eyes, &c., &c.

GLASGOW:
PRINTED FOR THE BOOKSELLERS.

EXPLANATION OF THE TREE OF FATE.

OBSERVE.—That you may either pick a number blindfolded amidst the leaves of this valuable tree, or throw for them with dice; if you pick for them and get among the branches, or in the blank leaves, it shows a speedy misfortune or disappointment at hand. The mark number of 1000 shows a great advancement in life, if you are so fortunate as to hit on it.

1 Gifts of Money	24 Visit to a foreign land
2 Prosperous run of business	25 Profit by industry
3 Speedy Marriage	26 Prosperity by marriage
4 Many Children	27 A multitude of cares
5 A good partner in marriage	28 By friends you will profit
6 You will become rich	29 Second partner better than first
7 Money through love	30 Surmount many difficulties
8 Cash by Trade	31 A false friend
9 A rise in Life	32 A pleasing surprise
10 A long journey	33 A change in your affairs
11 Anger and discontent	34 A ramble by moonlight
12 An important journey	35 [illegible]
13 A letter that will alter your present circumstances	36 Unpleasing tidings
14 Mind what you say to a lover	37 Loss in a short time
15 Present from a distance	38 A christening
16 Dispute with one you love	39 Get rich through a legacy
17 A law suit	40 [illegible] your situation
18 Visit from a distant friend	41 New wearing apparel
19 Party of pleasure	42 [illegible]
20 Preferment	43 [illegible]
21 Love at first sight	44 [illegible] in future
22 A prize worth having	45 You will be asked a question of
23 Wealth and dignity	importance to-morrow

20 Preferment	43 [illegible]
21 Love at first sight	44 [illegible] in future
22 A prize worth having	45 You will be asked a question of
23 Wealth and dignity	importance to-morrow

Figure 1: *The True Fortune Teller, or Universal Book of Fate* (1850)
Fate or free will? Chapbooks such as this showed that for many ordinary readers in the nineteenth century the dictates of fate were an important concern. In the later nineteenth century writers and reformers commented that books of fate and interpretations of dreams seemed to be amongst the most popular titles, competing with the romances and ballads that dominated the market. The potential outcomes listed here show most issues to be about money and love, with friendship and travel following close behind. There are no references to health, perhaps because of the huge literature already available specialising in health and quack medicines.

spectrum of human enquiry, from religion to genetics. This book explores the tension between them from a historical perspective that focusses on the relationship of time and modernity. It argues that each of these positions – on the one hand, a belief in the possibility of creating one's own life; and on the other hand, the belief that external conditions limit individual freedom – assumes different concepts of the future. In one, when agency is stressed, the future seems full of potential. In the other, when human action is limited by overarching structures or divine blueprints, the future is seen as more bleak. Social systems may not in themselves be a hindrance to progress, but they have often been portrayed as at odds with individual struggle, and therefore the source of at least short-term suffering. All of those in Bamford's autobiography who attempt to defy what seems to be their destiny meet with some kind of affliction.

Yet the tension between a belief in a preordained fate and the future as *tabula rasa* is not only a nineteenth-century phenomenon. It is still an important factor in political and theoretical debates about the comparative importance of individualism and social structure. To what extent can we be truly free to make choices about our lifestyle, aspirations, and experiences when class, ethnicity, gender, and global capitalism hedge us about with limitations and, more insidiously, with the discursive confines of our own culture? The struggle between these two ways of conceptualising social change – agency and structure – constitutes what one writer has called 'the contradiction which is at the heart of the dominant modern experience and which permeates our lives as a constant existential tension'.[3]

This book does not make any grand claim to resolve the agency vs structure debate. It does, however, point to one way in which, historically, a belief in agency has been strengthened. In the modern world we are all expected to be optimistic. The roots of this obligation lie in nineteenth-century Europe, whose culture embraced the credo that everyone should work towards improving their lives, a view which seems to reach its apogee in 1920 in the motto on the wall of the psychiatric clinic run by Émile Coué: 'Every day and in every way, I am getting better and better'. The immediate forebears of this philosophy were the nineteenth-century reform movements which represented free will as the individual's potential for self-improvement and an essential part of his or her contribution to the wider social good. A fundamental duty to improve can be seen in the work of, for example, the Society for the Suppression of Vice, which expressed its aims in terms of moral and social ideals, but was built on a belief in the desirability (if not the inevitability) of progress. Samuel Smiles's *Self-Help* (1858) is perhaps the most famous expression of this ideology, but a myriad of preachers up and down the country spread the same message.

Of course, to suggest the need for personal improvement is not necessarily an argument for free will, since it may be a matter of simply obeying natural laws, but the power of the individual to withstand vice and temptation was emphasised even by those who expressed a belief in social determinism. In fact, in a century when adherence to social and economic laws was often considered supremely important, several writers thought it necessary to argue that individuals were not the mere victims of these forces. The possibility that it might be 'society [that] prepares the crimes and the guilty person is only the instrument' was a frightening thought.[4] The answer, as Ian Hacking has shown in his historical study of statistics, was to stress the difference between determinism and fatalism. The future was not fixed; rather it could respond to human understanding of fixed laws. Certainly determinists taught that 'men's minds and circumstances ... are subject to laws as invariable as gravitation'; but this did not mean that 'circumstances' could not change. In the 1850s several writers forced a path through statistical determinism to suggest that human action also played a part. William Farr wrote:

> Introduce a system of ventilation into unventilated mines, you substitute one law of accidents for another ... As men have the power to change the conditions of life, and even to modify their race, they have the power to change the current of human actions within definite limits which statistics can determine.[5]

Despite the major nineteenth-century project of uncovering society's laws, then, individual agency would play an important part in human progress.

Progress, which still dominates the discourse of the so-called 'West', is the main concern of this study. Despite occasional prophecies of doom, progress, particularly in its technological form, is constantly held up to society as the justification for the advance of capitalism. The way in which this 'doctrine' of progress, during the period of its establishment, marginalised popular beliefs, especially those which represented the future as potentially bleak and foreboding, is the central motif of these chapters.

One way in which the cause of self-improvement was helped was in the labelling as 'superstition' of beliefs which did not advance that agenda. Progress required qualities such as independence, industriousness and honesty, which reformers suggested would all be undermined by the apathy promoted by superstition. This was a view put forward in the literature of social and moral reform, which, to judge from working-class autobiographies, was highly successful. Many self-educated artisans, looking back on the superstitious stories of their childhoods, bemoaned the uneducated belief in fate that had once been held by their contemporaries. Although he did not join in this disapproval, Bamford was typical of such a nineteenth-century autodidact, the working-class man who grasped the benefits of education, often against enormous odds. One of the gifts which such education provided, we are told again and again by those who struggled to achieve it, was the idea that each individual could forge their own future. Although the term 'agency' was not a contemporary one, the power of self-improvement was expressed in terms of achieving independence and respectability, criteria by which the struggling worker judged success.

Superstition

It is the contention of the following chapters that the struggle to confirm historical optimism as the foundation of social and economic endeavour in nineteenth-century Britain reveals the political agenda underpinning the definition of knowledge. In identifying superstition as undesirable, social reformers were not only demonstrating what they believed to be the difference between 'truth' and 'error', but were promoting a wish for the working classes to subscribe to a desire for self-improvement, a project that had economic implications, most obviously in the development of modern consumerism. The nineteenth-century struggle against superstition serves as an example of the fact that hegemony's first step is simply the placing of limits on what counts as knowledge.[6]

In the late eighteenth century the meaning of the term 'superstition' changed. Rather than conveying 'bad religion', as it had formerly done, it now suggested more a sense of 'misplaced assumptions about causality stemming from a faulty understanding of nature', as the *Encyclopaedia of Religion* describes it.[7] However, that faulty understanding was not solely linked to causality. It was also important to understand the workings of the natural world in order to be able to predict the future. 'Superstition', according to this new understanding of its meaning, actually made prediction more difficult, since it was premised on mistaken representations of the world. Accurate prediction was important – it lay at the heart of rationalism. Max Weber, foremost theorist of modernity, postulated that the principle of development inherent in the process of 'civilisation' was driven by the use of calculation as a strategy of social action.[8] Calculation, clearly, is dependent on the expectation of outcomes. Weber's concept of rationalisation, referring to the rise of a culture of planning, is, in fact, a form of secular prediction.

Those who promoted concepts of rational knowledge formulated an opposing concept of the 'irrational' which subsumed within it practices which may have had very little in common, other than the desire to know the future. The broad categories that are now used for forms of unorthodox belief within English-speaking cultures – superstition, the occult, magic – are a product of nineteenth-century reform. The arcane terminology of specific techniques (for example, scrying, geomancy, horary and genethliacal astrology) has been superseded. It was the desire to know the future, or the claim to be able to reveal it, that was the most important aspect of each belief as far as its opponents were concerned.

The phenomenon of prediction, then, once the preserve of magic and prophecy, has become crucially important to modern, secular society, but its change in meaning has not gone uncontested. This book argues that what futures studies have termed 'the colonisation of the future' lies at the heart of the conflict.[9] In appropriating the previously magical functions of prediction, modern forecasting has become a powerful means of excluding alternative interpretations of the future. The so-called decline of magic, in the West at least, has allowed a linear, exclusive teleology of progress to dominate. The case studies presented here will trace early stages of this process of exclusion to a genesis within the reform of popular culture in Britain, and will suggest that this was further developed in British colonial policy as the Empire found a rationalisation for its governance within a hegemonic definition of knowledge, one in which alternative conceptualisations of predicting the future could not be accommodated.

The reform of superstition was, in fact, a major player in the advance of individualism, and helped to mask the structural features which were likely to be determinants of the particular life. It is ironic that modern versions of many of these older popular beliefs have taken on board a stress on individualism, and now promote the virtues of personal responsibility in much the same way as those who disapproved of such beliefs once did. Late nineteenth-century influences, notably psychology and European fascination with Hinduism, have created a modern 'occult' which promises health, wealth and happiness if the subject thinks and acts in ways that transcend material conditions. The most extreme form of this 'New

Age' teaching even suggests that illness and disability result from inappropriate thought, either in this life or a previous one. In concentrating on individual 'error', these 'descendants' of magical tradition overlook the structural history of magic: its connections to class, gender, and culture.

The communal aspects of popular belief have been neglected. Owen Davies discusses witch-mobbing as an act of folk culture, and argues that it was a manifestation of extra-legal justice that has received little attention. It would not be carried out, he writes, unless sanctioned by a considerable portion of the community.[10] Similarly, many of the rituals described in the dream books discussed later in this study require the co-ordination of a group working together – for example, a number of young girls all in search of husbands. Going to see a fortune-teller was usually done with a number of friends, and it is certainly possible to interpret this as an active form of recreation and relaxation. This would not mean that those consulting the fortune-teller had no real belief in the efficacy of their undertaking, but simply that the shared nature of the outing was part of a sense of community that contributed to lightening the burden of their everyday lives. This interpretation draws attention to the roles of class and gender in the social function of magic, as a backdrop to the choice of groups with which to identify, in contrast with the more usual explanations of such activity as the mistaken gullibility of the weak and exploited. One account of beliefs amongst twentieth-century factory workers describes prediction as the resort of those who are excluded from power and influence, a means of allaying anxiety about a future over which they have no control. The desire to know the future, this common interpretation suggests, is the recourse of the powerless in a time of uncertainty.[11] Such an analysis underestimates the role of random misfortune in all lives, no matter how powerful they might be. Moreover, the superstition which nineteenth-century reformers so deplored in those who consulted a fortune-teller may actually have been grounded in a robust appreciation of life's limitations, which no amount of hard work or moral endeavour could overcome, and which turned the poor to activities which could provide a shared experience, perhaps even shared laughter.

It has been suggested that the very idea of modern, secular prediction helps to create community by excluding the possibility of difference:

> [T]he act of prediction exhibits a way of thinking that is limited to a certain cultural understanding. Moreover, the act of prediction is part of a cultural worldview that constricts the future to being only one future, by assuming that it is already 'out there' in some sense, waiting to be discovered.[12]

Some non-Western scholars have queried this 'extrapolation-of-the-present' scenario, particularly since the extrapolation is generally done by those whose present is the Western, capitalist world. Such privilege has political repercussions:

> 'Imagining a future', taken literally, does not mean to think about the future in some undefined sense – it means to create an image of what the future will look like. And how will we do this? ... [A] notion of a future is a notion that creates meaning. A picture of the future is a picture with meaning.[13]

The stories told here will reveal individuals who did not expect the future to be a more dynamic continuation of the present, but rather feared decay, decline, and catastrophe. They problematise the restricting 'extrapolation' scenario, and furnish a reminder of how very recent a phenomenon it is.

This book builds on *Visions of the Future*, in which I analysed the rise of the rational almanac, that compendium of statistics and data of which *Whitaker's* became the archetypal example. I argued there that the reasons for the disappearance of the old astrological almanacs which had preceded *Whitaker's* could be traced to middle-class disapproval of 'superstitious' prediction, and that this was in part based on a different understanding of time. The marginalisation of superstition had previously been discussed largely in terms of the rise of scientific knowledge, the spread of education, and increasing technological mastery of the environment, all interpretations that were fundamentally premised on the assumption that the superstitions were wrong, errors to which no educated person could subscribe. In particular, it has been suggested that the growth of towns and cities contributed to the disappearance of old beliefs.[14] Keith Thomas's ground-breaking study of religion and magic in the early modern period places a strong emphasis on the importance of urbanisation as an enlightening factor, arguing that among the reasons for the disappearance of magic from a central and intellectually important position were factors such as literacy and access to the printed word, which promoted the spread of rational ideas.[15]

Rather than beginning from the premise that truth will inevitably triumph over error, then, this book examines the disapproval of popular belief from the point of view of how such disapproval privileged its advocates, with the expectation that this will expose conflict over differing truth systems.

Progress

Comparison with other temporal practices has in the past played an important role in defining one's own culture. For example, the level of accuracy which was achieved by British astronomers in the nineteenth century contributed to the formation of a national identity that was often expressed in disapproval of 'primitive' or uneducated understandings of time. Much early anthropological literature consisted of the observations made by Western ethnographers about the time concepts of other peoples precisely because time-consciousness was considered such an important part of 'civilised' culture. Johannes Fabian's important book, *Time and the Other*, has discussed how the discipline of anthropology itself was premised on the supposition that the observer and the observed understood time in different ways.[16] Recent scholarship has investigated examples of the political use of temporal 'knowledge', from the Spanish conquest of the Aztec empire to the economic disadvantage of Puerto Ricans in New York.[17] If we accept a politics of time, it becomes apparent that cultural constructions of temporality have often provided elites with a vocabulary for 'substantiating the legitimacy of their rule'.[18] In particular, Carol J. Greenhouse finds a common pattern amongst those claiming dominance, in which they represent themselves as embodying progress. The notion of progress is indeed closely allied to hegemony, since it is defined by those in power: the change they bring is, of course, portrayed as a change for the better.

The belief that the passage of time brings improvement can be found throughout history, but there have been particular periods in which this belief has seemed to receive more attention.[19] It is an idea associated above all with nineteenth-century Britain. The *Oxford English Dictionary* lists 1852 as the first appearance of a characteristically Victorian meaning of the 'future': 'a condition in time to come different (especially in a favourable sense) from the present'.[20] J. B. Bury's influential study of 1920 suggested three phases in the development of the idea of progress: the first period up to the French Revolution, when 'it had been treated rather casually'; the second, following the establishment of scientific positivism, when it was a familiar idea in Europe, but 'not yet universally accepted'; and the third, post-Darwin, which 'established the reign of the idea of Progress' (enthroned with a capital P).[21] In fact, the whole 'Enlightenment project' was based on the fundamental belief that history would deliver increasing understanding and control of the natural world. As one recent writer has put it: 'If there is a core promise the modern world had to keep, it was that reason leads to progress. Reason applied to organising life gives history meaning by making it a story of the overcoming of tyranny and want.'[22]

Generally, discussion of the question of progress has hinged on the idea's accuracy or otherwise. Is humanity likely to experience 'a gradual ... increase of goodness, or happiness, or enlightenment'?[23] Or are we rather unleashing forces which we cannot control, and which are likely to lead to decline, social unrest, and even global catastrophe? In the wake of the twentieth century, older pronouncements of progress may seem to have been naively optimistic. Indeed, some scholars have identified the disappearance of a belief in progress as a hallmark of 'post' modernity.[24]

Rather than simply dismissing earlier belief in social progress as mistaken, however, it might be useful to think about why it was once so widely accepted and still remains part of the implicit underpinning of much of contemporary life.[25] A great deal has been written about its development in the eighteenth and nineteenth centuries; but in an important and innovative analysis, *The Idea of Progress in Eighteenth-Century Britain*, David Spadafora writes that this body of scholarship is 'a badly flawed torso'. He believes that existing studies concentrate too much on elite ideas, with little social context, and he identifies a tendency 'merely to relate who believed in progress during a given period and to record the words into which those individuals put their belief'.[26] It is common, for example, to find reference to Auguste Comte, who, in the mid-nineteenth century, suggested that it was the rise of scientific procedure in his 'positive philosophy', that is, in sociology, which had allowed progress as the true grand design of history to be identified.[27] In contrast to this concentration on philosophical and elite historiography, Spadafora pays great attention to social context, and in order to do so coins a useful distinction between the *idea* of progress and various *doctrines* of progress, the former term conveying a belief in improvement or change in a desirable direction, while the latter signifies the expression given to that belief in a particular social group, such as writers on education, or Christian eschatologists.

This study addresses the shortcomings of the 'flawed torso' in a way that is different from but complementary to Spadafora's. It suggests that the idea of progress was constructed in part by a dialectic between elite philosophy, such as that of Comte, and plebeian custom. It sets out to recapture some popular attitudes to time – beliefs of those who were excluded from wealth, power, and status – and it argues that a strong component of such beliefs was the fear that life might become even more difficult. This pessimism was markedly different from an elite desire to portray society as full of promise for individuals who were morally deserving and worked hard. Prosecutions for fortune-telling, usually presented in historical accounts as a simple matter of the social control of vagrancy, were actually part of a wider story of temporal conflict, with middle-class reformers keen to control representations of the future by discouraging the fatalism implicit in popular prediction.

Spadafora writes that Victorian optimism was 'a complicated and fragile mood', which seems to have been 'neither singly focused nor uniformly expressed'; yet, he says, there has been no attempt to determine 'whether at any time there were identifiable social groups or intellectual contexts in which the belief in progress customarily appeared'.[28] Later chapters will suggest that one specific context in which such a belief could be found was that of the social reform movements which sprang up to tackle public and private disorder in the new urbanised centres of industrial Britain.

In the late 1970s and the 1980s several historical studies appeared in what might be labelled 'the reform of popular culture' school. These argued that in the late seventeenth and the eighteenth centuries the ruling classes, fearful of the potentially disruptive forces at work amongst the 'people' at large, increasingly demarcated the bounds of their own good taste and gentility, criticising practices and attitudes which they felt were unacceptable. Perhaps the most influential historian to have advanced this thesis is Peter Burke, whose 1978 book, *Popular Culture in Early Modern Europe,* was one of the first to propose the model;[29] but there soon followed several other studies which extended the same argument to the nineteenth century, examining the 'reform' of popular recreations, customs, and sports. These demonstrated that a multitude of customary practices (bear-baiting, street football, fairs) became subject to public criticism and attempts by municipal authorities and government to ban them.

This model of reform can be extended to the question of time. Time became a focus of social tension, with the newly emerging middle classes subscribing to beliefs in punctuality, orderliness, and diligence which they often found lacking in the lower orders. As a result, the language of social reform was frequently couched in temporal terms, criticising laziness and advocating the profitable ordering of the day. Clocks and watches, 'the glamour technology of the eighteenth century', were eulogised as expressions of order and dependability.[30] God himself was pictured as the cosmic watchmaker, and the universe as his creation was said to follow immutable, rational laws, like clockwork.[31]

Very often the history of time is relayed as a narrative about technological development. However, time is not only about mechanical precision or scientific realities. Each culture takes the physical phenomena – the passage of the sun, the sequence of the seasons – and tabulates and names them for a variety of different purposes. The nature of time is not simply 'there', waiting to be discovered. It is a component of cultural communication, part of the discourse through which we structure our experiences in order to convey them to each other and to ourselves. Each society formulates its own understandings of different kinds of temporality, and, in fact, contains varieties of temporal understanding that differ from the dominant interpretations, sometimes consciously articulated as forms of resistance, sometimes subverted and shameful.

In nineteenth-century Britain there were popular beliefs about time that ran counter to the ideal of clockwork regularity. One such belief was recorded by the writer Mortimer Collins in 1869. He was experiencing a high attrition rate amongst the bees he kept in his garden, and when he mentioned this to a local labourer, the man asked if there had been a death in the family: 'bees would infallibly die after the death of anyone who cared about them, unless they were told of the event'. This prompted thoughts from a neighbour in whose house a kitchen clock had refused to go since a recent death; and yet another declared that 'on the death of a relation of hers a clock which had been stopped for thirty years revived and struck the whole twelve hours'. Collins was less bewildered about the beliefs regarding bees – somehow any mysteries about their behaviour seemed natural and 'immemorial'. However, strange beliefs about the relationship of deaths to clocks appeared to him to be puzzling 'in a parish where the three R's are sedulously taught'. Why should clocks stop – or stopped clocks go on again – when people die?[32] The possibility of technological efficiency being influenced in some non-material way by human developments seemed thoroughly incongruous, and although Collins writes with interest about these beliefs, he labels them unambiguously 'superstition'.

Although these stories of Berkshire villagers were couched in terms of clocks, it is clear that it was not mechanical failure that was at issue, but rather something much more intangible, and that time itself seemed to respond to human death. This is not simply some antiquarian curiosity. As Collins pointed out, the existence of such a belief went to the heart of the educational programmes inspired by nineteenth-century rationality; and it was not only rationality that was challenged.

Wherever those who hold one belief dismiss another as superstition, it is certain that the interconnectedness between knowledge and power may be seen. Only a few years after Collins's encounter with these beliefs, the world drew close to a 'unification of time' when an international conference voted to recognise the meridian passing through Greenwich as the prime meridian. Time, surely, had become the most all-pervasive manifestation of modernity, and efficiency in measuring it marked Britain out as the most modern of nations. British authorities were, of course, happy to accept the symbolism of Greenwich as prime meridian. The word 'temporal', after all, has associations not only of time but also of secular power of the most material kind, as distinct from the spiritual realm of the church. Possessing the standard by which all watches and clocks throughout the Christian world were to be measured contributed to the establishment of empire in a myriad of ways. However, just as a counter-imperial discourse was always present alongside the imperial, even at the very zenith of British power, so a counter-temporal discourse can be detected alongside the apparent homogeneity of Greenwich Mean Time.

If time could be experienced in different ways in different places, the authority of British temporal measurement might seem all too local. When time could stand still for a Berkshire clock, what might it do in the most distant regions of empire, where 'natives' might as yet be unaware of its existence? Whether in a colonial context or in reference to the working classes within Britain, 'alternative' temporalities were indeed identified as dangerous, and attempts were made to contain or disempower them. Temporal heterogeneity was attacked both within Britain and in the colonies as part of the attempt to establish central control and efficient communication.

A history of heterodox time can, therefore, be seen as part of a post-colonial analysis of the nineteenth century. Writers using post-colonial insights have done a great deal to highlight temporal politics. For example, Heather Goodall writes that the people at Ernabella, South Australia, thought that the British nuclear bomb tests of the 1950s caused the measles epidemic of 1948 which killed at least a quarter of their people within two weeks. Goodall does not dismiss this explanation as simply 'wrong', but explores the connections and meaning for the Anangu people. 'In doing so she pits western understandings of chronology and causation against Anangu understandings of colonialism, disease, and death.'[33] In discussing rationality in the discourse of history, Dipesh Chakrabarty has stressed that 'minority' history is not just a matter of being included, but of methodological and epistemological questioning of what the very business of writing history is all about.[34]

It would generally be true to say that those writing with post-colonial methodologies have posited a dominant nineteenth-century idea of progress which has faced challenge since the post-colonial turn. Anne McLintock, for example, in an influential study of colonial race, gender, and sexuality, argues that 'the almost ritualistic ubiquity of "post" words in current culture ... signals ... a widespread, epochal crisis in the idea of linear, historical progress'.[35] Yet resistance to progress has not been only in the 'post' era; alternative temporalities were present throughout the eighteenth and nineteenth centuries.[36]

As in *Visions of the Future*, which depended chiefly on nineteenth-century almanacs, the major sources used here are literary, but in this case they are fortune-telling and dream books, as well as works of fiction. Each of the chapters describes a conflict at the meeting point of world-views, and each centres on the question of temporality. In total they show advances being made by the idea of progress, and help to provide detail in what has been a broad-brush account of historical optimism.

The first chapter deals with the familiar story of the rise of centralised time. The increasing importance of clock time, the co-ordination of railway company schedules, the establishment of the prime meridian: all these are well-known signposts in what appears to be the inexorable advance of temporal efficiency. This chapter highlights, however, the extent to which this 'advance' was dependent upon a particular understanding of the future. Indeed, modernity has been characterised as the need to conceptualise and generalise the future.[37] The blank calendars of the late nineteenth century gave very material form to the idea of progress by replacing the crowded historical chronologies of the traditional almanac with spaces on which to record the precise schedules created by the railroad and steamship companies, as well as the many appointments that a busy, modern life would entail. At the same time, the new calendars also symbolised a *tabula rasa* of opportunity, on which individual members of society could write their futures, increasing the emphasis on individual responsibility.

The second chapter examines one of the most visible of the groups which chose to target the fundamental pessimism of popular prediction. The case which the Society for the Suppression of Vice brought against the astrologer Joseph Powell demonstrated fear of the disruptive potential of prediction, as well as disapproval of an array of 'unrespectable' behaviours associated with attempting to influence the future, such as gambling on the lottery. In unravelling the implications of the charges brought against Powell, this chapter shows how respectable fears of 'superstition' were associated with disapproval of a particular understanding of forward planning.

The third chapter looks at the popular chapbook genre of dream books, and argues that this widespread tool for prediction was appropriated at the end of the century by psychoanalysis, when Freud's own 'dream book' revolutionised attitudes to dreams. Like gambling, the interpretation of dreams was a feature of popular culture which claimed to give access to a better future. In the case of dream books this was through the increased knowledge and understanding which they promised their readers. If you knew in advance what your future husband would look like, for example, it would be easy to resist the advances of inappropriate suitors and to wait patiently for destiny to unfold. In finding success for his new, rational lexicography of dreaming, Freud ensured that the power of dreams became focussed on understanding the past, increasing, again, the individual's responsibility to create the future.

The fourth chapter explores the theme of the future's links with national identity. It takes a specific example of the cross-cultural political significance of temporality, examining the use of the word 'Dreamtime' as applied by British anthropologists to the beliefs of indigenous peoples in Australia. The introduction of the word 'Dreamtime' in the 1890s to describe Aboriginal beliefs should be considered in the light of the long tradition of unrespectability and unrest associated with visionary dreaming in English culture, and in the light of the escalating discussion of the true nature of dreaming. The use of this particular translation reinforced a representation of indigenous culture's position as remote from the enlightened understanding of science and progress. It signalled both a marginality and a threat of disturbance. Aboriginal temporal 'otherness' was presented as a timelessness, a total unawareness of time, which was not only unable to measure time accurately, but was also seemingly confused about the distinctions between the past and the present. Such apparent errors, so the dominant discourse would have it, marked Aboriginal society out as particularly in need of temporal reform. Teaching the virtues of punctuality, regularity, and dependability became the project of missionary and administrator alike. This experience was not confined to Australia. Wherever European explorers and colonists travelled, they took the same temporal certainties about punctuality and progress. Even today, temporal certainties lead scholars to write about other cultures' lack of concern with exactness in time: Navajo, Arabs, Latin Americans.[38] Embedded in this construction of 'lack' lie the same assumptions about other cultures' lack of forward planning. The intellectual high ground of this position has important economic implications, particularly at a time when Western science and medicine negotiate their relationship with non-European modes of thought in a global economy that increasingly requires cross-cultural contact.[39]

As the calendar approached the turn of the twentieth century, the essentially British nature of faith in the future was conveyed in the popular press. *Punch* began to use the image of the female body as an icon of hope that was closely associated with the imperial destiny. Chapter 5 examines how, emerging from earlier uses of the figures of Britannia and Marianne, *Punch*'s young women appeared in its annual calendar in full-page illustrations welcoming the new year. These images not only conveyed a sense of progress, they also underscored that broadcasting a message of confidence in the future was part of the national destiny. It was part of the national character. This chapter also examines how 'faith in the future' marks Joseph Conrad's Lord Jim out as 'one of us'. It is Jim's 'judicious foresight' and forward planning that are the most important reasons to overlook his failures and accept him as a hero.

It will be clear from the above summary that the chief sources for this study are records of educated commentary on popular culture, often disapproving commentary. In most cases we cannot move beyond these to approach some unmediated primary record of plebeian belief. The clearest example of how dangerous this exercise might be, how far it might lead us from the goal of recreating *mentalités* with anything approaching accuracy, has been revealed by Robert Poole in his excellent unravelling of accounts of the famous calendar riots of 1752. These riots, so the usual story goes, were a result of the uneducated masses believing that the change from Julian to Gregorian calendar, in which the year jumped from 2 September to 14 September, actually deprived them of eleven days of the year. 'Give us our eleven days!' has been an illustration of just how stupid the mob could be, in accounts ranging from Hogarth in the eighteenth century to recent twentieth-century books on the calendar. Poole demonstrates convincingly that these riots probably never took place, and that references to them reveal rather a 'characterization of the common sort as "the mob", with an intellectually debased collective identity ... The construction of the calendar rioters, then, was part of a wider process of cultural stereotyping by which the superstitious and ignorant mob had imputed to it qualities the reverse of those claimed by the enlightened.'[40] The image of the confused peasant is, indeed, 'one of the most misleading and the most eloquent of historical myths',[41] in the sense in which myth conveys both something which is fictitious, but also something which embodies a deep, archetypal aspect of culture. Whether or not Mortimer Collins's neighbouring villagers really believed that deaths could stop clocks is perhaps not the central concern of this present study. What we can say with some certainty is that many middle-class commentators believed that large numbers of the working classes held irrational and odd views on the subject of time. In setting out to change these views, social reformers helped to create a national identity that was fundamentally premised on a sense of temporal superiority.

The role of time measurement in the creation of modern industrial society has been examined from several different perspectives: triumphalist accounts celebrating technological efficiency, anthropological studies making cross-cultural observations, and social histories about labour campaigns to limit the working day. Perhaps the most influential of all analyses has been Edward Thompson's article 'Time, Work-Discipline and Industrial

Capitalism', which argued that in the late eighteenth and early nineteenth centuries the controls imposed by clock schedules were internalised by workers, who came to look on the possession of a watch as a mark of status.[42] Here I am following where Thompson led, by arguing that another type of internalisation took place in the nineteenth century, not of temporal accuracy but of temporal uniformity: the expectation that what was 'right' at the metropolitan centre was right for all localities and even across the colonies. Those who administered the British Empire were perfectly conversant with longitudinal variations in the measurement of the day, but there were criteria by which temporal comparisons could be made other than simply the hour on the clock. One of these criteria was a doctrine of rational progress dependent on present-centredness. Certainly the future was to be planned, and a whole array of official activities was devoted to that enterprise (the subject of another book than this). However, the future was not the domain of supersitious prediction, with all its potential to distract workers from present efforts.

The widely held assumption that magic and superstition were replaced by the progress of enlightenment fails to appreciate the extent to which the secularisation of prediction was also a social history of the changing management of time. This history belongs to the story of the rise of individualism, one of the dominant motifs of the society we live in today. We are still taught in our schools that hard work will lead to rewards. Psychological testing 'proves' that children who are able to cope with deferred gratification turn into more competent, stronger individuals.[43] The Protestant ethic is alive and well. The structural limitations on what individual effort can achieve have been much obscured, to the benefit of capitalist production. An older belief in fate may point us not merely to the survival of superstition but also to a very salutary grasp of the limits of individualism and a truer assessment of the structures at work in creating all our futures.

1
Clocks, Calendars and Centralisation

In 1884 Greenwich was declared by an international conference in Washington to be the prime meridian of the globe. As the word 'meridian' suggests, this indicated that in the ordering of space and time, Greenwich was in effect the centre of the world.[1] The choice of London as the standard by which other times and places located themselves was brought about by respect for British efficiency in the measurement of time: throughout the nineteenth century, the chronometers of the Greenwich Observatory were the most accurate timepieces in the world.[2] The decision also indicated how important the *Nautical Almanac*, which used Greenwich longitude as the basis for calculations, had become to mariners worldwide. This dominance of London as a temporal centre obviously had repercussions on colonial and foreign policy, but its significance lay not only in international affairs. It can also be examined for what it reveals about domestic culture. Throughout the nineteenth century urbanisation and industrialisation exerted an accelerating impetus on British society, with Victorians uncomfortably aware that they were living in an age when speed was increasingly important. A growing tyranny of schedules and appointments and an anxious sense of needing to manage one's time were characteristics of the age. This chapter briefly traces the development of a sense of time based on London, with particular consideration of what this suggests about changing understandings of the future. It argues in particular that the obligation to exhibit foresight, to plan sensibly, was widely considered to be an integral part of the proper use of time, which in turn was portrayed as the responsibility of each individual. Centralisation on the national capital, London, and growing individualism came together within the field of temporal practices.

As Britain's prosperity came to depend more and more on trade, rapid systems of transport were devised, for which co-ordination was

Figure 2: *Greenwich Observatory, seen from Croom's Hill* (about 1680)
From an oil painting by an unidentified artist, in the National Maritime Museum.

necessary. Road transport underwent dramatic changes between the mid-eighteenth and mid-nineteenth centuries, to the extent of being called a 'revolution' by at least one historian.[3] The Royal Mail was an important force for change. Losing customers to stage coaches, the Post Office was keen to adopt new designs which allowed speed to be combined with comfort. By 1820, according to one estimate, the competition between passenger and mail coach had resulted in a situation in which it was quicker to travel by coach than on horseback.[4] This competition in terms of speed led naturally to an increasing emphasis on schedules and predictability of arrival times. One result was to introduce 'a new note of exactitude' into business transactions, since communication between cities could now be planned ahead with a fair degree of certainty. The journey between London and Birmingham, for example, could generally be relied on to take only a couple of days.[5]

The emphasis on speed and precision was carried further by the rail networks that spread across Britain in the middle of the nineteenth century. From the opening of the Stockton to Darlington railway in 1825, through the 'boom' years of the 1850s and 1860s, to the era of suburban commuter travel in the 1920s, the huge expansion of the railway system transformed the landscape of Britain. It brought with it an increasing dependence on timetables. In 1840 the Great Western Railway began to use London time in all its timetables, and in 1846 the North Western Railway introduced London time at their Manchester and Liverpool termini.[6] *Bradshaw's Railway Timetable* appeared for the first time in 1839 and there followed a decade of what its critics portrayed as temporal confusion, with *Bradshaw's* timetables using local times while railway companies increasingly referred to London time. *Punch* thought *Bradshaw* a promising target for satire, and in a 'guide to the guide' referred to its 'Deal Time', 'Irish Time', 'Dinner Time', and passengers whose 'Trains do not start for some time'. Nevertheless, for all its shortcomings, 'Few literary efforts ... have ever achieved so much for the cause of Progress as has the Book of BRADSHAW.'[7] When, in 1848, even *Bradshaw* began to use standard London time for all train timetables, London's ascendancy was recognised in a book which *Punch* said was consulted by 'the Queen, the Prince of Wales, both Houses of Parliament, and all the Government offices'.[8] Small wonder, in an age of secularisation, that most stations displayed a prominent clock, often on a tower, just as the village church had once done: a suitable symbol for the replacement of the man of God by the travelling businessman.

The Clock

There can have been few more visible symbols of time observance in Britain than the time-ball on the roof of the Greenwich Observatory, itself sitting prominently at the crest of a hill. George Airy, the Astronomer Royal, installed a time-ball in 1852, in what was to provide the model for several other centres. When the master clock within the observatory reached 1 p.m. every day, an electrical impulse dropped the ball from the top of the rod to the bottom, in a visual display of scientific accuracy and temporal authority. Significantly, those clocks around the country which by arrangement took their time from Greenwich were called 'slave' clocks.

The power to decide what the correct time was, and to send that message around the world, was enormously important. In Joseph Conrad's *A Secret Agent*, set in 1886, anarchists plan to blow up the Royal Observatory. They view their attack as an assault on science, which they say the powerful regard as 'in some mysterious way ... the source of their material prosperity'.[9] Moreover, the plotters tell each other, 'the whole civilized world has heard of Greenwich'. The plot of Conrad's story was inspired by a real event of 1894, the 'Greenwich Bomb Outrage', in which a young man was found in Greenwich Park, horribly mutilated after having, apparently accidentally, set off explosives which he himself was carrying. Conrad, in common with the press of the time, assumed that an attack on the observatory had been intended. Although he dismissed such an intention as 'perverse unreason', in fact an attempt on the bastion of time measurement was not surprising. After the establishment in 1889 of Paris time as the standard for the whole of France, France continued to run 11 minutes 20 seconds earlier than Greenwich, in what has been described as a tacit acknowledgement of the political undesirability of using a British measurement as a standard.[10]

Links between Greenwich and British influence can be drawn even earlier than world recognition of the prime meridian. In order to calculate longitude correctly, British ships had been dependent since 1767 on the government's *Nautical Almanac*, first drawn up by the Astronomer Royal, Sir Nevil Maskelyne. It was this almanac, using tables of lunar distance from the sun, which began the use of

Greenwich as the starting point for all navigational calculations. In order to be able to find their position at sea, navigators would calculate the angular distance between the moon and the sun, then compare the local time with that at which the same lunar distance was said by the tables to occur at Greenwich. Finding the difference between the time at the place of calculation and the time at Greenwich allowed temporal distance to be transposed into spatial difference, to be expressed in terms of longitude east or west. In other words, one's position on the globe was calculated according to a relationship to Greenwich.

The story of the search for navigational safety and its links with the search for a method of calculating longitude has been told often. A mistaken assessment of location was the chief danger at sea, and although in practice this was not always a matter of longitude, accuracy of longitudinal calculation became emblematic, symbolising the goal of human mastery over the dangers of the natural world. In 1714 Parliament passed *A Bill for Providing a Publick Reward for such Person or Persons as shall Discover the Longitude at Sea*, and in order to decide the allocation of the said reward, of £20,000 at most, a board of 23 commissioners was set up.

Although Nevil Maskelyne's tables of lunar distance revolutionised the calculation of longitude, and indeed made him a likely candidate for the reward, it has been another method, that of developing a timepiece that could be set at Greenwich time and carried on board ship, that has most captured the modern imagination. The clocks that were developed in response to the 1714 Longitude Act have been the object of admiration for their style, their beauty, and the story of the heroic struggle that produced them. The goal was to produce a clock that would keep such dependable time that it could be transported across the globe on journeys lasting several years, moving across temperature zones and areas of great humidity, yet still be relied on to give accurate Greenwich time. This could then be compared with local time, calculated from the stars, in exactly the same way as Maskelyne's lunar distance method, giving a comparison that could be transposed into longitudinal difference. John 'Longitude' Harrison, a watchmaker of Barrow, Lincolnshire, set out to produce such a timepiece. The story of his struggle to perfect his craftsmanship, and to convince the Board of Longitude of his entitlement to the £20,000 prize, has been a popular one, especially when seen as a heroic conflict of a man of the people against the elite machinations of the unprincipled Astronomer Royal, jealous of a reduction in status of his own innovation, the tables of lunar distance.[11]

Whatever the truth about rivalry or bad feeling between Maskelyne and Harrison, it would be fair to say that it was Maskelyne's *Nautical Almanac* which laid the foundations for the rise of Greenwich, and therefore for the elevation of London, as the arbiter of spatial and temporal location. The new almanac rapidly became an indispensable accompaniment to any sea journey: it enabled navigators to cut the time needed to calculate longitude from over four hours to about half an hour, and it provided tables for four years ahead. Until the appearance of Maskelyne's almanac, it had been usual to express longitude as a number of degrees and minutes east or west of the point of departure or destination, but now a measurement based on Greenwich became almost universal. One estimate suggests that by the end of the nineteenth century approximately three-quarters of the world's shipping was using charts based on Greenwich.[12]

While navigational developments were assuring the dominance of Greenwich time, parallel developments on land were to achieve a similar result within Britain. The dominance of 'Greenwich time' was advanced by developments in the 1850s in which the observatory made use of the latest technology, electrically driven clocks and telegraph cable, to facilitate increasingly wide communication of its timekeeping. Airy's time-ball was part of this programme, with the dropping of the ball connected to time-balls all over the country. A time-gun was installed at Dover Castle, and was fired at noon by an electrical current communicated direct from Greenwich. The symbolism of the sun at its zenith combined with the sound of the gun from castle ramparts could hardly have been a more stirring reminder of the custodianship of time at the heart of the Empire. A proud George Airy commented in 1865: 'The practical result of the system will be knowledged by all those who have travelled abroad. We can, on an English railway, always obtain correct time, but not so on a French or German railway, where the clocks are often found considerably in error.'[13] His comments are an unambiguous demonstration of the close links between achievements in the measurement of time and national pride.

From 1859, the most visible clock of all dominated London itself. Big Ben, erected in the tower of the new Houses of Parliament, was the apogee of clockmaking expertise.[14] Almost a hundred years later, Lord Reith sang its praises, in describing the sound of its chimes on radio. The sound of Big Ben, he wrote, helped to bring rural areas 'into direct contact with great Empire institutions like Parliament and the Royal Observatory at Greenwich'.[15]

A legal decision on how to resolve the anomalies between local times and London time was frequently sought mid-century, and was finally introduced and carried in 1880. Parliament declared that:

> Whenever any expression of time occurs in any Acts of Parliament, deed, or other legal instrument, the time referred shall, unless it is otherwise specifically stated, be held in the case of Great Britain to be Greenwich mean time, and in the case of Ireland, Dublin mean time.[16]

The story of how Greenwich Time was later superseded by Co-ordinated Universal Time (UTC), which came to be the standard time around the world, has been told elsewhere.[17] In the twentieth century, telegraph, telephone, and radio signal allowed increasing global co-ordination. Time, based on atomic clocks, is now co-ordinated from Paris. The temporal dominance of Britain did not outlast its Empire.

The Calendar

Centralised time was not only a matter of clocks and schedules; it had long been a major impetus for calendar reform. Perhaps the most significant development in Britain was the adoption in 1752 of the Gregorian calendar, replacing the Julian version, a change which brought Britain into line with her nearest European neighbours. Robert Poole's important study of this 'reform' suggests that secularisation was a major factor. He believes that there was a shift amongst the educated classes towards seeing the calendar purely in astronomical terms, rather than ecclesiastical:

> The reform of the calendar, based as it appeared to be on improved techniques of astronomical measurement ... fitted this framework [of progress] very nicely ... progress was now the flow of time itself. An accurate calendar was a psychologically important foundation for such a world-view ... In a sense, the reformed calendar was the foundation stone of the temple of national enlightenment, the altar raised over the idol of superstition. The calendar reform,

then, was in all these ways symbolic of the progress of enlightenment against popular superstition.[18]

The cultural changes signalled by the 1752 reform were given
further impetus early in the nineteenth century by the abolition of
the stamp duty on calendars. Until then, by far the best known
versions of calendars were those called almanacs (because of long
association with astronomy and astrology), which were published
by the Stationers' Company in London.[19] The stamp duty had made
calendar publication prohibitively expensive for other companies,
but when it was repealed, the market was flooded with hundreds of
new titles. These new calendars, freed from the constraints of
centuries-long tradition perpetuated by the Stationers' Company,
can be used as a litmus test for changes in the way in which people
recorded the passage of the year.

More than clocks or watches, which were as yet to find a
technology which would make them cheap enough to be widely
available, calendars remained the most widespread guide to the
passage of time in the middle of the century. However, as their
format changed, the guidance they gave also altered. It had been
common to find the term 'remembrancer' in their titles. Amongst
those things for which they served as a reminder were historical
events, of which a list of significant dates was given in the
chronology; but the almanac had also traditionally included a strong
emphasis on the ecclesiastical structure of the year, with reminders
about saints' days and church anniversaries. It was this characteristic which particularly offended one radical reviewer of some of the
new titles, writing in 1848:

> It is painful to observe in the present day, when Almanacks are
> sold by thousands, and read by millions, that in the majority of
> them, so many remnants of superstition are allowed to remain in
> the shape of saints' and holy-days ... The influence of the constant
> recurrence of the names of these days must be pernicious –
> implanting an idea of periodical sanctity in minds not given to
> much thought.[20]

The ecclesiastical year could be an important organising principle.
For example, while travelling to Australia on board the *James Pattison*
in 1834, Mary Yates Bussell felt that time on the long journey had
been suspended. The way she used to recreate for herself some sense

of its passing was to think about what the date was in England: 'It was passion week in England', she writes in her diary, as if it were not also for her on board ship. Real life, and real time, went on 'at home' in England, and the most easily remembered aspect of that time was her place in the ecclesiastical year.[21] Within rural England, similar attitudes were noted:

> An old man will thus tell you the dates of every fair and feast in all the villages and little towns ten or fifteen miles round about. He quite ignores the modern system of reckoning time, going by the ancient ecclesiastical calendar and the moon.[22]

The role of remembrancer, however, was gradually replaced. In 1859 Parliament decided that certain dates would no longer be officially commemorated in the church calendar as set out in the Prayer Book. The Acts for the observance of the 30 January, 29 May, 5 November, and 23 October were repealed. 30 January was the day on which the death of Charles I had been commemorated; 29 May the anniversary thanksgivings instituted by Charles II to mark his Restoration (suitably timed to coincide with his birthday); and 5 November the day on which public thanksgiving had been the practice since the Gunpowder Plot of 1605. 23 October was a date instituted by the Parliament of Ireland in 1662 to commemorate the defeat of the Kilkenny uprising in 1641, a Catholic movement resisting Protestant settlement of Ulster. This clearing away of remembrances helped to transform the calendar from a dense eighteenth-century text to what we would recognise now as a collection of largely empty spaces to use for appointments. The transition from early modern practice is described by Poole:

> No-one who has studied early modern almanacs and calendars can fail to be struck by the sheer complexity and variety of cycles of time woven into them. The modern diary, with its blank ordinal of dates, punctuated by the odd random public holiday, is a meagre thing by comparison. Every day (like every sunset) was different, according to whether it was a red-letter or a black-letter day, a feast or a fast, a Sunday or a weekday, a dog day or a winter day, a first day or a quarter day, a day in the high spiritual realms of Easter or in the long drag between trinity and advent, a day when the moon was waxing or waning, a day for sowing crops or

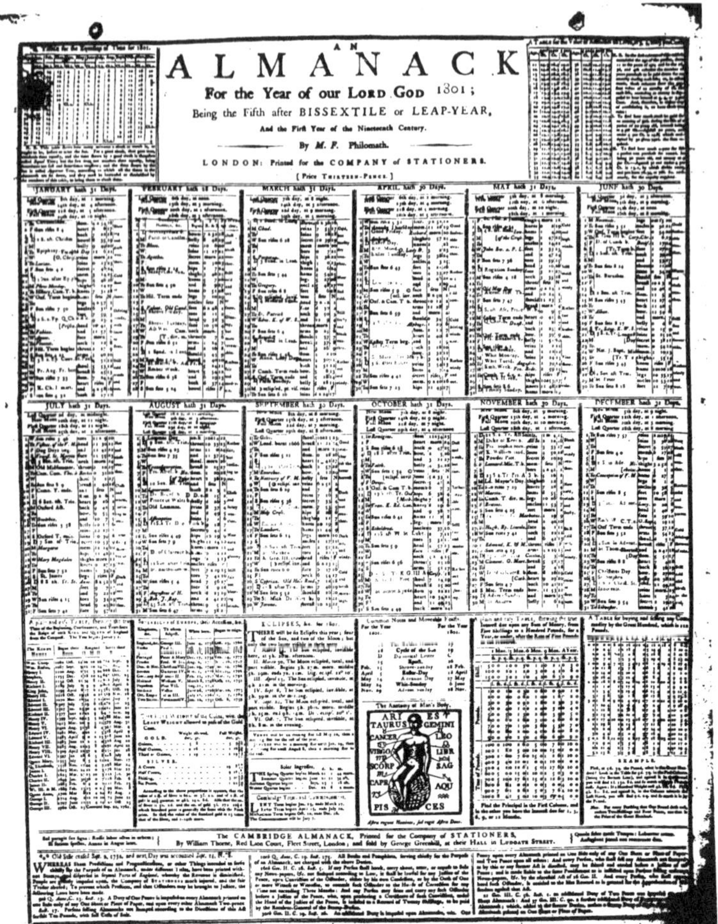

Figure 3: *The Cambridge Almanack* (1801)

This Stationers' Company broadsheet is an example of the crowded nature of the pre-nineteenth-century calendar. Within each month is a column which gives the moon's rulership over parts of the body. This is related to 'Zodiac Man', the drawing of the human body with astrological signs. It was thought that the sign in which the moon was on a particular day influenced the health of the part of the body ruled by that sign, and this information would have been useful for the application of bleeding, herbal medicine, etc. The calendar also includes a column of weather predictions based on planetary aspects. There were several 'county' almanacs produced by the Stationers, as well as hundreds of local calendars produced cheaply and illegally. All these were replaced as the century went on by increasing commodity advertising and by the kind of blank diary exemplified in Figure 4.

for letting blood – and, after 1752, whether it was an old-style feast or a new.[23]

In contrast with these earlier calendars, the modern calendar became increasingly a blank slate, representing the future rather than a commemoration of the past. It is this Lockean promise of a future on which any intentions may be inscribed that is the key to understanding changes in the role of calendars during the nineteenth century, and to appreciating their value as a window onto cultural change.

There may have been many reasons for the clearing away of what were thought to be anachronistic remembrances in the calendar. Parliamentary discussion preceding the 1859 changes shows proponents of the reform citing an increasing distaste for the anti-Catholic sentiment that such anniversaries might encourage. Earl Stanhope read from the church services for 5 November and 29 May and pointed out that the wording 'implied that there was some connection between the tenets of Roman Catholics and the practice of assassination'.[24] While not defending the 'errors and corruptions' of the Roman Catholic Church, the earl was bound to say that in the present day no one would regard all Catholics as conspirators actuated by 'hellish malice'. Although he did not refer to the 1829 Act restoring civil liberties to Roman Catholics, the context of increasing religious toleration in nineteenth-century Britain was clearly a factor in the calendar changes. The Archbishop of Canterbury supported the call to end divisive practices, saying that it was totally undesirable for any sentiment to be spoken in prayer which was 'likely to call up feelings of indignation in the breasts of ... fellow-countrymen'. Again, a shared time was, by implication, a component of a shared nationhood.

Criticism of the calendar for perpetuating social division had occurred earlier in the century, when the publisher Charles Knight had condemned *Merlinus Liberatus*, one of the Stationers' almanacs, for dating each issue from 'our Deliverance by King William from Popery and Arbitrary Government'. This reference to the 'Glorious Revolution' of 1688 clearly marked the original author of the almanac, John Partridge, as a Whig, and his political position was perpetuated for the entire life of the almanac, long past his death in 1715, until the last issue in 1871. The front cover also regularly commemorated the Jacobite plot of 1695, with 1801, for example, being 'the 106th year from the Horrid, Popish, Jacobite Plot'. Knight criticised the almanac for 'keeping alive, for the purpose of exciting religious animosity, the memory of transactions which are a disgrace to the character of this country, and the worst blot upon the history of its law'.[25]

In fact, Knight himself was no Tory, but his desire for social unity, expressed in the early years of the century when fears of revolution were being voiced by some, was an expression of a desire for shared knowledge as a basis for cultural consensus. Literacy, education, self-improvement, all are familiar themes of nineteenth-century social reform. All rest in part on a shared sense of time, both of the past and the future. In the 1860s *Wing's* almanac, the second-longest-running of all almanacs after *Moore's*, began to note the royal family's birthdays for the first time in the body of the calendar, and to print them as red-letter days like ecclesiastical holidays.[26] No doubt the Stationers intended such universal commemoration to be a unifying role, counteracting earlier criticism of sectarianism and divisiveness.

As well as the desire for a spirit of national unity discussed in Parliament, the pursuit of industrial productivity also contributed to calendar changes. Ronald Hutton relates how, between 1790 and 1840, employers 'carried out a ruthless pruning of the Christmas holidays':

> In 1797 the Customs and Excise Office, for example, closed between 21 December (St Thomas's Day) and 6 January (the Epiphany) on all of the seven dates specified by the Edwardian and Elizabethan Protestant calendars. In 1838 it was open on all except Christmas Day itself. The Factory Act of 1833 put the seal upon this process by declaring that Christmas and Good Friday were the only two days of the year, excepting Sundays, upon which workers had a statutory right to be absent from their duties.[27]

As opposed to dates actually printed in calendars as official holidays, the practice of taking holidays varied from region to region, as shown, for example, in the study of Lancashire wakes by John Walton and Robert Poole.[28] Douglas Reid has examined the practice of 'Saint Monday', that is taking a holiday on the day after the weekend, a purely customary practice that was opposed by employers with varying degrees of success.[29] One historian has commented on widely differing policies:

A mine at Levant in Cornwall where an old custom of giving six holidays a year was observed was mentioned as exceptional in 1842, and of another mine in the same district it was stated that whereas formerly there were many holidays in the year, there were now only two recognized holidays, Christmas Day and Good Friday.[30]

At a different social level, but responding to similar economic imperatives, the Bank of England closed on 47 holidays in 1761, 18 in 1830, and only 4 in 1834: Good Friday, Christmas Day, 1 May, and 1 November.[31] The trend towards reduced holidays became a subject for discussion in debates leading up to the Bank Holiday Act of 1871, which established Boxing Day, Easter Monday, Whit Monday, and the first Monday of August as national holidays. The Act signalled the increasing secularisation of society, as Parliament discussed the need for holidays purely to ensure the well-being of workers and the efficient conduct of business rather than to commemorate a religious observance. The individualisation of the calendar was also accelerated towards the end of the century, as annual holidays became increasingly a matter for negotiation between employer and employee. Increasing opportunities for travel by rail made the holiday a popular development, but employers were at liberty to allow this or not as they chose.

Another aspect of nineteenth-century calendar change was a decrease in local content. Again, Poole comments about the early modern calendar: 'On top of [seasonal and ecclesiastical variations] were days of particular local significance: a wake, a civic event, a fair, a rent day, an assizes day.'[32] Some regional publishers had defied the Stationers' Company's dominance even before 1834 in order to produce an almanac with local appeal that won purchasers away from the London product. Birmingham, Newcastle, and Manchester almanacs, for example, were particularly well known. The Stationers recognised this aspect of the market, and themselves issued several county almanacs: the *Cambridge, New London, Middlesex, Cornwall, Gloucester, Norfolk, Warwick, Cheshire, Wiltshire, Yorkshire,* and the *Shrewsbury.* These were discontinued during the early years of the nineteenth century, a further sign of the increasing centralisation of time consciousness. New titles, both from the Stationers and from their rivals, tried to appeal to purchasers in a general way, regardless of where they lived. Although regional almanacs did flourish in the northern textile districts of Lancashire and Yorkshire later in the century, they were something of a novelty, praised and copied by antiquarians for their apparent value in perpetuating dying traditions. By then, the norm was taken to be the bland new rational almanac, devoid of local colour and distributed across the country by an expanding rail network. The archetype of this new 'universal' calendar was *Whitaker's*, first published in 1869.

Urbanisation played a part in these changes. *Rider's British Merlin* contained a list of all the fairs in Britain, with dates and locations. Such a calendar would have been of most use to tradesmen and workmen plying their wares across the country in rural areas and small towns, and was of little interest in urban centres. One satirical reviewer of the 1802 edition of *Rider's* suggested that Londoners had no need to know when the sun rose (as shown in one column in the almanac) 'except it be a hint to break up company', and suggested more practical advice than 'that perpetual care which Mr. Cardanus Rider advises us to take of our cabbages'.[33] The last edition of *Rider's* was in 1828.

In the early twentieth century, calendar reform was increasingly associated with attempts to perfect a universal calendar that would help to unite humanity in a common system of measurement of the year. The early inspiration for this project was the 1849 publication of Auguste Comte's *Positivist Calendar of 558 Worthies of all ages and nations*;[34] Comte dated all years from the French Revolution of 1789, embracing that revolution's attempt to remove the calendar from ecclesiastical influence and to place it entirely 'on rational grounds'. In pursuit of this goal, he abandoned all 'relation to the moon', and made each month invariably four weeks long, with each year of thirteen months. All months would, therefore, begin predictably with the same day of the week. Most notably, all months and festivals would be renamed in honour of 'Worthies of all ages and nations', heroes of rationalism such as Descartes, Fichte, and Hegel.

The League of Nations was active in the early twentieth century, with aims very similar to Comte's. Its general conference on communications in October 1931 was in fact the first international conference on calendar reform.[35] For those who aspired to world peace, the universality of the experience of time was a promising source of commonality. In practice, however, this could be interpreted as the universal relevance of British values. In *Vo-Key's Royal Pocket Index*, a small pamphlet produced in London in 1901, the 'Greatest Discovery ever made on the subject of universal time' is propounded: the fact that the hour hands of all watches ('when they have the proper time') point simultaneously to all the different times of the world at once. This was a simple matter of taking the hour to which the watch hand was pointing as standing for the Greenwich meridian, and then using the clock face to represent longitudinal divisions, so that hours before or after Greenwich could be calculated. However, the writer of *Vo-Key* is thrilled that with a little practice, 'the eye will be trained to see the meridians of all places', and lists among many possibilities Sheffield, Oldham, Bristol, Chicago, Cadiz, Santiago, and Samoa.[36] The world could indeed be thought to centre on London.

The Future

There were two distinct ways in which the space once taken up by dense lists of anniversaries and chronologies was reinvented. An increasing proportion of the calendar was devoted to the new practice of advertising; and blank space was left to accommodate the practice of writing in appointments. Both these developments were connected with planning for the future; one with buying goods to secure comfort and health, the other with the conservation of time. Letts, which became a major publishing name in calendars and diaries during the nineteenth century, produced new 'future-focussed' diaries.[37] The publication of the first Letts diary in 1812 originated 'a new concept of diary-keeping completely different from the traditional use as a personal historical record', or, indeed, different from the calendar dominated by considerations of the ecclesiastical year. This new diary, intended for recording the movements of ships to and from the Port of London and for recording finances based on bills of exchange, was predominantly a commercial diary. It made this clear by including only the days of the business week, omitting Sunday altogether.

Almanacs had always had something of a commercial role. When Francis Moore published his almanac at the end of the seventeenth century, he used it to publicise the benefits of the health pills he was producing. When Henry Andrews continued with *Moore's* at the end of the eighteenth century and the beginning of the nineteenth, he announced to his readers that they could buy barometers of superior quality and reliability, made by himself. Compilers often included their address for correspondence, so that readers who wanted to purchase an astrologer's advice could consult them in a letter. Some

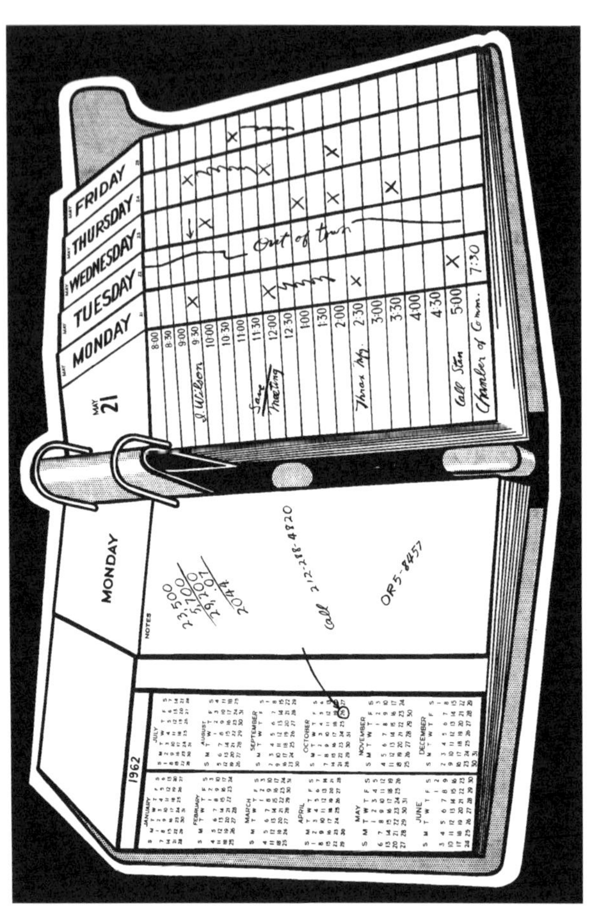

Figure 4: *Calendar*, Roy Lichtenstein, 1962. Oil on canvas. 48½ × 70 inches.
The Museum of Contemporary Art, Los Angeles; The Panza Collection

This modern image contrasts sharply with the *Cambridge Almanack* of 1801. Here the blank space allows for appointments, planning, personal accounting. Note especially that the only days allowed for are the working week. The ecclesiastical functions of the calendar have disappeared altogether.

advertised their teaching services, offering mathematics and astronomy. What so many authors of almanacs called a 'service' – the provision of the important data of the year – furnished an opportunity to sell other services. This association between calendars and advice (on health, the future, husbandry) was made clear when the almanac duty was removed and the market was quickly inundated with titles whose primary purpose was to advertise a product, whether that was material (health pills, tailoring, or stays) or spiritual (commitment to evangelical belief, or a life of temperance). There did not need to be an identifiable advertisement as such; simply the name of the publisher in the title of the almanac was often considered sufficient:

> The free distribution of almanacs which combined vestigial calendars ... with advertisements and price lists became one of the major categories of the dramatic expansion of business mail following the introduction of the Penny Post in 1840. Items such as *L. Hyam's Tailoring, Fashionable Ready Made Clothing and Outfitting Almanack for 1846* were brought to post offices by the sackful.[38]

The absence of what we might now easily recognise as an advertisement was attributable to the advertisement tax, an amount varying from 1s 6d to 3s 6d at various times during the nineteenth century. Until this was abolished in 1853, the promotion of products within almanacs was usually carried as text, indistinguishable from other material about the weather, husbandry, or chronology. Since the same tax was payable whether the advertisement consisted of two lines or several pages, those who could afford to pay the tax at all naturally invested in as much copy as they could. In 1844 *Rogers'* dialect almanac carried advertisements for the first time, 17 pages of them altogether, with 7 entirely for Parr's Life Pills. In 1845 it carried 15 pages advertising Worsdell's Pills. There was much contemporary disapproval of the 'quacks' whose products were so widely advertised, but almanacs had a long tradition, through medical astrology and herbalism, of dealing with matters of health.

Even after the abolition of the advertisement tax, book almanacs continued to carry large numbers of advertisements. In the 1860s, for example, *Bradshaw's* attracted attention by the sheer number of its pages devoted to advertising. The first 32 pages, one reviewer noted, were all advertisements.[39] Letts also carried advertisements for the first time in the 1860s. With no advertising duty, it became increasingly possible for advertisements to enter the consumer's home as part of the common sheet calendar, now in effect little more than a handbill. This, combined with decreased costs of illustrations and a rise in the middle-class standard of living, contributed to the late nineteenth-century boom in advertising, allowing the printed image to extend its area of influence.

Although middle-class standards of living were improving, overseas competition increasingly threatened market growth, and it has been suggested that there was a decline in business confidence in the third quarter of the nineteenth century.[40] One measure to which families could resort for protection against declining fortunes, if they could afford it, was insurance. As Lori Anne Loeb has noted, many advertisements traded on middle-class fears, and one of the products most frequently canvassed in calendars and diaries was insurance. This was a form of commercial activity that was undergoing tremendous expansion at the end of the century.[41] Cassell's 'date blocks' were advertised as being specifically suitable for insurance companies, on the assumption that diaries and calendars were essential to the efficient records of an insurance company as it kept track of transactions and predicted future risk. Sometimes the language of insurance was used in consumer advertising, underlining the increasing exploitation of anxiety in the home. A Cocoatina advertisement asked: 'Do you give your children Cocoatina? If not you place them at a disadvantage in the struggle for life with 1,000,000 children who use it daily. Eleven pence is a trifling premium to pay on a permanent insurance of health, wealth and beauty.'[42]

Many diaries included an insurance policy in their purchase price. This provision reflected the use of a diary to plan travelling, since it was travel insurance above all that was thought to be exercising the mind of the purchaser. The Letts diaries for 1901 all insured against railway accident. One, the 'Cyclist Pocket Case', had two prices; one for twelve months covered risks due to collision only (1s) and one was for 'all risks' (1s 6d).[43]

An advertisement was designed to encourage the consumer to buy future security, but the blank space that was to play an even more important role in modern calendars was also intended to appease anxiety. To plan the future was to be in control. In such planning, portability was an important feature. No matter where you were, you never needed to be unaware of time, and indeed you could save valuable time by having the calendar reference close at hand. In 1844 *The Times* recommended calendars for all the different parts of a house: 'No. 1 The Nursery. No. 2 the Dressing Room. No. 3 The Kitchen. No. 4 The Cupboard. No. 5 The Garden. No. 6 The Stable.'[44] These were clearly no longer works of literature, as Mortimer Collins had once described the calendar, providing leisurely reading for the entire year, but rather pointed reminders about immediate circumstances, serving as a focus for activity and thought.

In fact, the saving of time was seen as a type of insurance. Letts published a whole series of small pocket books which included the *Ladies' Register of Calls*, 'a concise record of calls received and paid', which would show at a glance who would be 'at home today', thus saving wasted visits. The *Bachelor's Washing Book* provided counterfoils for laundry, allowing the busy bachelor to keep an easy check on whether his clothing was returned. The purchase of consumer goods, no doubt a factor contributing to the frantic pace of life, could also be chronicled, with a 'Gentleman's Cellar Book', in which to 'keep an exact account of all wines and spirits laid down and used', a 'Garden Account Book', an 'Inventory Book', and other titles such as a 'Register of Letters Despatched'.[45] All these together served the purposes of address books, account books, birthday books, debt books, rental books, and appointment diaries, functions which had once been filled by the almanac alone. A glance at many eighteenth-century almanacs reveals pencilled margin notes or pages inserted at the end of the almanac, listing loans, debts, purchases, birthdays, and so on, all those commitments for which Letts now produced individualised reminders. Such a proliferation of personal accounting testifies to wide cultural changes, but the title under which Letts chose to display this list speaks of one particular agenda. All the above were advertised under the heading 'Charles Letts's Time Saving Publications'. The saving of time so advocated was in some senses a type of investment. By investing the time it took to keep your lists of expenditure or purchases or social visits, you would later save time spent in a muddled search for what you should already know, in the kind of silly fuss so perfectly exemplified by Lewis Carroll's White Rabbit, who rushes around looking for the gloves he has carelessly put down somewhere. It is exactly this attitude to timesaving which is demonstrated by modern calendars, which offered themselves as a means to more efficient personal organisation. The 'timesaving' offered was, in fact, a matter of record keeping, so that time was saved by increasing organisation and increased planning. Indeed, records of time could be life saving, as demonstrated in Wilkie Collins's highly popular novel *The Moonstone*, in which the whole denouement of the plot hinges on the exact date of a journey undertaken by Lady Glyde. This important detail can be provided by the daughter of the narrator, Gabriel Betteridge. He writes:

> The only difficulty is to fetch out the dates, in the first place. This Penelope offers to do for me by looking into her own diary, which she was taught to keep when she was at school, and which she has gone on keeping ever since.[46]

The underlying message of this happy outcome is to illustrate how right social reformers have been to stress the educational importance of literacy and timekeeping. It is Penelope's ability to write (gained despite her father's lowly status as a house-steward) which provides the crucial legal evidence, but, more importantly, it is her awareness of the need to keep an account of how she has spent her time that saves the heroine from wrongful incarceration as a madwoman and allows truth to prevail.

The temporal regulation of work can be seen reaching an apogee in the development of Taylorism in the 1890s, when 'scientific management' attempted to squeeze the most productivity out of the present moment.[47] But it was in the increasing number of offices that calendar organisation was most essential. The earliest record we have of offices keeping diaries is the order of the Prime Minister, the Duke of Wellington, that *The British Almanac* should be made available in all government offices after 1828. Early the next century, the Clerk of the Stationers' Company received a letter from Cassell's reporting that '"Whitaker" has established itself as the universal book for offices and business men generally'.[48] He added that *Whitaker's* sales were in turn being affected by the new 'Daily Mail Year Book', which had such enormous resources in the way of publicity and distribution that it was 'absolutely unassailable'. The growth of office time can be seen in the Post Office's objection in 1888 to the suggestion that the 10 a.m. signal from Greenwich be dropped: 'We cannot dispense with the 10 o'clock current from Greenwich ... Every office throughout the United Kingdom is supplied with this time current by a system which has taken many years to establish.'[49]

A growing sense of rushing around, of being pressed for time, has long seemed an entirely appropriate way of depicting a nineteenth-century society in thrall to railroads, steam engines, and the clock. Many accounts relate comments on the increasing speed with which life seemed to be conducted, 'the number of things crowded into a day, and the rush from one to another'.[50] Often this sense of 'busyness' was self-congratulatory, as in Jeremy Bentham's enthusiastic eulogy of the modern: 'The age we live in is a busy age; an age in which knowledge is rapidly advancing towards perfection. In the natural world, in particular, every thing teems with discovery and with improvement.'[51] Yet in a brilliant study of the cultural implications of time consciousness, Stephen Kern refers to England as having 'the least expectation of a radically new future' of any of the combatants in the Great War.[52] He quotes A. J. P. Taylor's observation: 'In England landowners granted leases for 99 or even 999 years, equally confident that society and money would remain exactly the same throughout that time.'[53] In fact, there is no real contradiction between the portrayal of a nation increasingly obsessed with the future and one quite sure that the future would resemble nothing so much as the present. British national identity was premised on a certainty, however unconscious, that the unexpected would not happen. This was actually a prerequisite for the culture of planning. It was to have important repercussions on how British society treated those who expressed different attitudes to the future.

2
Fortune-Telling

Prediction of the future is always a contested area, involving as it does authority and status. Belief in a particular prediction implies trust, and the trust of the working classes was a contentious issue in nineteenth-century struggles over superstition. Reformers expressed a special responsibility to protect the newly literate, who had begun to read all sorts of different publications, but whose naive gullibility, it was thought, exposed them to the problem of which information could be believed. This question of the potential corruption of the innocent was central to the status of popular belief. Condemnation of magic rested on the fear of exploitation. This chapter examines how the marginalisation of superstition involved implicit questions of danger, and most especially of sexual danger.

One notable example of the uncertainty about authoritative knowledge was the confusion over weather prediction that took place in the 1860s. Robert Fitzroy, the first head of the new government department of meteorology, decided to issue weather forecasts to the newspapers, but these proved disastrously wrong. On the receiving end of parliamentary criticism and journalistic scorn, he committed suicide in 1865.[1] In the short term, this unhappy episode made meteorology departments highly sensitive about the nature of the terminology they used. They avoided the words 'prognosticate' or 'prophesy', terms that were associated with the kind of prediction that might be made by an astrologer or cunning man. However, the term 'forecast', coined by Fitzroy, was also tainted for several years following his failure and death. In 1872 the British meteorology department began to issue a daily weather 'report', not a forecast, and in the 1870s the American Federal Meteorological Service began to discuss weather 'probabilities', before changing to 'indications'. Even in 1879, a contributor to the journal *Notes and Queries* wrote about weather 'forecasts', using quotation marks to convey his own awareness of the problematic nature of the word.[2] No-one used the term 'prediction'. This was far

Figure 5: 'Mabel Wray and her Companion Consult the Astrologer',
The Fortune Teller (1870)

There is little doubt about the subject which concerns Mabel Wray as she consults this wizard. The ceiling has suspended from it the figure of a dragon, magical symbol of sexuality, and it hangs directly over Mabel's head. The gloom of the surroundings suggests the occult atmosphere into which astrologers withdrew in order to avoid being prosecuted under the vagrancy laws. This is a very different sort of astrological practice from that of William Lilly in the seventeenth century, whose fame allowed him to claim that he was 'familiar to all'.

too closely connected with practices which educated thought dismissed as superstitious.

Predictions were most closely associated with astrologers and fortune-tellers. Since 1736 belief in both had been treated by the law as the sole preserve of the gullible and the foolish. In that year the Witchcraft Act repealed earlier Jacobean legislation against witches, on the assumption that witchcraft simply did not exist. There were new penalties, however, and these were against those who 'pretend to exercise or use any kind of witchcraft, sorcery, inchantment [sic], or conjuration, or undertake to tell fortunes'.[3] This, in itself, was seen as a sign of progress, since Parliament was suggesting that no educated person could any longer believe in such superstitious practices. The new crime was part of the Vagrancy Act, and, through several amended versions, Parliament suggested that those who practised the various categories of prediction were likely to be idle and disorderly, found in the ranks of rogues and vagabonds. The description of behaviour by which such rascals were to be recognised varied each time the legislation was brought up to date. In 1822 those using 'any subtle Craft' were included in a list of offences that gave prominence to 'common Stage Players' and 'those performing an Entertainment or any Part or Parts therein'. In 1824 the question of charging money for predictions was mentioned for the first time, accentuating the assumption of fraud in the fortune-teller's practice. Indeed, in 1824 fortune-telling was actually the leading item in what had become an exceedingly long list of offences likely to be committed by vagrants. It took pride of place ahead of obscene exposure, betting in the street, carrying a cutlass with an intent to commit a felonious act, and 'sundry other behaviours' likely to cause a disturbance.[4]

The 'enlightened' attitudes exemplified in this legislation have been credited with a general disappearance of 'magical' practices in modern Britain, particularly in urban centres. The most widely influential interpretation of the decline of magic is that offered by Keith Thomas, who argues that among the reasons for the displacement of the magical world-view were factors such as literacy and access to the printed word, which were connected with urbanisation.[5] His analysis implies that there was little magic left to investigate by the beginning of the nineteenth century: '[M]en emancipated themselves from these magical beliefs.'[6] However, if fortune-telling was indeed in decline, why was it placed first in the offences listed by the 1824 version of the Act? In 1744 it had come ninth, well after 'all fencers and bearwards' or 'common players of interludes'.[7] We might, perhaps, deduce from this early nineteenth-century prominence in the legislation that 'magical' beliefs were actually widespread and annoyingly recalcitrant.

Magic did not disappear altogether. Astrology and fortune-telling are still with us today, having negotiated the Victorian labyrinth of disapproval, scorn, and legal control. True, their modern manifestations are different in important ways from earlier practices, but they nevertheless maintain some sense of connection with earlier forms. Much of what Hugh Cunningham writes about continuities in nineteenth-century leisure applies also to popular belief.[8] Like those sports and pastimes which Cunningham finds continued throughout the nineteenth century, despite considerable attempts by reformers to ban them altogether, magical practices survived in altered form. Indeed, magic itself might be seen as a form of leisure, 'something that unbends the mind by turning it off from care'.[9] Consultations of 'magicians' were often undertaken in groups, and might end in laughter. The only recorded instance of Samuel Bamford visiting a fortune-teller was when, while working for a landowner at Prestwich, he accompanied the cook and the housemaid, both young women, to see 'a famous seer' known by the name of Limping Billy. Bamford stayed at the local public house waiting for them, but he got impatient and tried to slip into the room to join the group, whereupon he was shouted at by the 'wizard' and told to get out. Afterwards all three young people adjourned to the tavern:

> [W]hilst there partaking a glass of warm liquor, they told me that old Billy had caught me laughing, and was very angry at my daring to laugh in his presence. I admitted that I certainly had been betrayed into a not very reverential feeling when I saw them listening so demurely whilst the old imposter peered into his dirty cup and mumbled his prognostics ... I ridiculed old Bill and his trade until they joined me in laughing at their adventure as well as my own, and so in this lively mood we set off towards home, and arrived there better pleased with ourselves and our journey than we had at one time expected to be.[10]

The anger of Bamford's Limping Billy might easily have been caused by the fear that the newcomer was an informer likely to report him to the law, a development which could have led to imprisonment or transportation. What follows in this chapter is an example of just such a turn of events: the arrest and trial of an astrologer with evidence brought by someone who had pretended to consult him. The story reveals the processes marginalising popular belief at the beginning of the nineteenth century, a demonstration of the interconnections between respectable knowledge and respectable morality, and the way in which the development of the latter led to an increasing withdrawal of magical practices from the public domain. Several more clients of a 'wizard' can be seen engaged in a 're-creation', seeking reassurance about important decisions at a time of rapid change and increasing pressures.

On 10 October 1807 a woman selling matches from door to door called at a prosperous house in Great Russell Street, Bloomsbury, the home of Mr Blair, a surgeon. When the maidservant, a girl called Ann Mintridge, answered the door, the match-woman pressed into her hand an advertising bill, which offered the services of a professor of sidereal science – an astrologer. Ann put the handbill down on a sideboard, and later the master of the house picked it up. Knowing that fortune-telling was illegal (and astrology was counted as a form of fortune-telling), he decided that he would set out to capture the rogue whose services were advertised. Giving his manservant, Thomas Barnes, a half-crown piece, he sent him to the address on the bill, together with a letter, which he himself had written but which he pretended was from Barnes, requesting the nativity, or birth chart, of Ann Mintridge, as well as asking questions about whether she would marry, whether she would have children, and so on. After collecting the completed chart and interpretation the following day, Barnes went straight out to notify a waiting constable, who then arrested the astrologer, Joseph Powell, and took him into custody.

The charge brought against Powell was under the Vagrancy Act. He himself seemed to accept that this was the main issue, when, after the magistrate had chastised him for his idle ways, he burst out in a fervent affirmation of his willingness to work. He could call many people to testify that he had worked for them, he declared. 'I have worked in gardens; I have cleaned house; I have cleaned windows; I have beat carpets; I have scoured rooms.' He had even offered to sweep the street after the dung cart, but they refused to take him, saying that he would not be able to stand up to the demands of the work.

The defendant's words survive because the Society for the Suppression of Vice, which brought the prosecution, aroused strong criticism in its pursuit of this case and felt obliged to publish its own account. Mr Blair, the surgeon, belonged to the Society, and although it had at one time made a decision not to defend its members in their crusading activities, it chose in this instance to support him with the strongest possible action. This was because reports appeared in the London press which were highly critical of the prosecution, and which, according to the Society, resulted in a

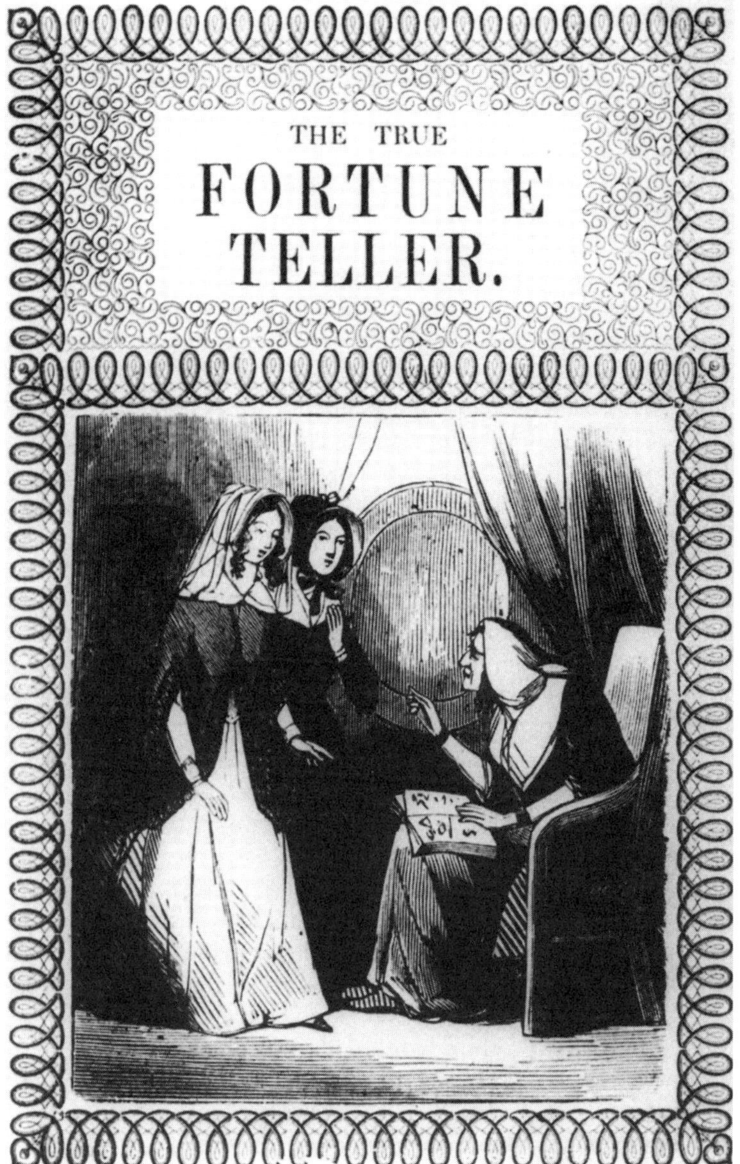

Figure 6: *The True Fortune Teller* (1850)
Another couple consults the fortune-teller.

'torrent' of abusive letters sent to Blair's house. Several newspapers portrayed the astrologer as a man more suffering than sinning. 'Every line of his countenance seemed a furrow of grief and anguish', the *Times* reporter wrote. 'Aged, tall, meagre, ragged, filthy, and care-worn, his squalid looks expressed the various features of want and sorrow.'[11] In contrast, Blair was painted as one who had been presented with the handbill by his *factotum*, as a '*morceau of novelty*, for his amusement'.

The Society for the Suppression of Vice had experienced opposition ever since its foundation. Formed in 1802, it was one of the many reform societies that sprang up in England and Scotland in the late eighteenth and early nineteenth centuries, responding to a sense of concern about accelerating social change. The titles of these societies were often self-explanatory; for example, the Society for Bettering the Conditions of the Poor, and the Society for the Suppression of Public Begging. They found their support largely within the new urban middle classes, and for that reason did not always have a very wide base. The Society for the Suppression of Vice itself was famously dubbed by one of its critics, Sydney Smith, as 'the Society for suppressing the vices of persons whose income does not exceed £500 per annum'.[12] Some criticism came from unexpected quarters, such as from within the ranks of other middle-class reform groups. This may have been because the Society decided to limit its membership to orthodox 'Churchmen', thus offending dissenters and 'puzzling' many evangelicals.[13] William Cobbett also accused it of '[giving] the laws ... an extension and a force which it never was intended they should have'.[14] The use of entrapment was particularly criticised, a strategy it used in sending Barnes to consult Powell.

The Society's pamphlet, *The Trial of Joseph Powell*, was published in 1808, and takes issue with the 'flagrant misrepresentation' of the defendant as presented in the press. Contrary to the picture of Powell as suffering 'want and sorrow', the Society claimed that his astrological practice at 5 Sutton Street, Soho Square, had been thriving for about 30 years. More than 10,000 handbills for distribution around the streets had been printed by only one of his printers (and he gave work to several others). In court, the prosecution claimed that letters seized from his house proved that he had three or four wives, making the point that he was gaining enough money from his fraud to support several dependants. One letter of his was read out which had been written to a woman bemoaning the fact that he was losing £200 a year through being distracted from work by his passion for her. His clients, too, were not poor. One, specifically mentioned by the prosecution, though not named, was 'female, not of the lowest class', and another was referred to as 'a lady residing near Holborn'.[15]

The fact that Powell was described as having been in business for 30 years immediately rings warning bells. If he had practised lucratively and openly from his own home for so long, why had the law not been enforced at some earlier date? One explanation for an uneven history in prosecution has been put forward by Owen Davies in his study of witchcraft and magic. He posits a period, following the Witchcraft Act of 1736, of 'conscious elite detachment from the beliefs of the "people"', succeeded by a renascent interest towards the end of the century when there was a growing perception of superstition as a problem. Such a perception, he suggests, was partly due to the rise of Methodism: 'Increasing anxiety over the popularity of Methodism led to a closer examination of the spiritual and moral state of the labouring poor, particularly by members of the Church of England, whose religious interests were threatened.'[16] Indeed, Methodist clergy were at times thought more likely to believe in witchcraft and diabolic possession, and to be prepared to help out with exorcism and prayers in such cases.[17] David Vincent suggests a gendered dimension to this difference between denominations, implying an increasing masculinisation of church culture: 'It is perhaps no accident that the only Nonconformist sect seriously to entertain the reality of witches in the early nineteenth century, the Primitive Methodists, was also the only one to give genuine responsibility to its women members, and as witchcraft was later expelled from its theology, so the ranks of lay preachers became a predominantly male preserve.'[18] Although the Vice Society was later joined by several prominent evangelicals, its early campaigns against superstition may have reflected this lack of Dissenters in its membership.

The reasons given by the Society itself to justify its actions provide a means of analysing why Powell was prosecuted in 1807 rather than at some earlier time. It was particularly a London-based organisation, which repeatedly voiced concern about the dangers of a rapidly increasing urban population, especially the potential for public disorder. It expressed a desire to protect innocent new arrivals to the city who might be duped by Powell's claims, and led by him into further criminal behaviour. It certainly did not feel that magic was disappearing. According to its own record, in its first 15 years it brought 39 cases against fortune-tellers, as compared with

11 prosecutions of keepers of disorderly houses. Later in the century, it became best known for its campaign against pornography and prostitution, so that in this early stage of its work, to prosecute more fortune-tellers than brothel owners signifies an importance attached to the perceived threat of magical practices.[19] This is urbanisation seen not as the harbinger of educated rationality but as the threat of lawlessness.

One concern caused by urbanisation lay in problems about the boundary between public and private. In the early years of the nineteenth century the Vagrancy Act became the focus of these anxieties, when increasing dissatisfaction was expressed about the power of London magistrates to interpret its application.[20] When the new offence of indecent exposure was added in the 1820s, the newspapers gave voice to what they claimed to be widespread public outrage. Constables took into custody several people who seemed to the press to be undeserving of arrest: some courting couples who were caught kissing in the shadows, and elderly men 'relieving themselves' after dark. Significantly, the newspapers which expressed a wish to preserve a 'traditional' sense of individual liberties, and which were highly critical of the enforcement of the Act, were *The Times* and the *Morning Chronicle*, exactly the same newspapers mentioned by the Society as being opposed to the conduct of the Powell case.

The nineteenth-century city seemed, to contemporaries at least, to enforce new levels of visibility. There was some resentment at claims that it might be necessary to enter under false pretences into someone's home to see what their misdemeanours were, or to crawl through the bushes in public parks in order to spring out on young couples, as some inspectors were accused of doing. One prominent critic of the Vagrancy Act, the lawyer John Adolphus, was especially scathing about the offence of 'threatening to run away', which could be used against any man with wife and children. The legislation was clearly responding to fears that the deserted families would be left as a burden on the rates levied for the Poor Laws; but as Adolphus commented:

It does seem most hard and unjust to punish ... a man for a hasty word or inconsiderate expression, followed by no act whatever. There is a ferocity in such a penalty which ... is the more galling as it exposes every man to punishment on the assertion of any discontented or spiteful individual.[21]

In 1824 there was a parliamentary investigation into the workings of the Vagrancy Act, and the subsequent report showed that arrest and detention on this charge were not at all unusual.[22] Men really were fined or imprisoned simply for threatening to leave their wives. As Adolphus said, an Englishman's home was no longer his castle. New definitions of public and private action were being asserted.

The Society for the Suppression of Vice displayed a wide-ranging and controversial interpretation of what constituted the public sphere. Its use of an *agent provocateur* was one of the aspects of the Powell case that particularly angered the London press. However, the Society felt justified in this tactic because it believed that private morality and public order were closely connected. Although it did not bring any charges against Powell on the grounds of his affairs with women, despite accusing him of bigamy, it drew clear connections between his fortune-telling practices and his immoral life. Superstition seemed at the very least to give opportunity for, if not actually to cause, private vices.

When called to report to the Police Committee of the House of Commons in 1817, the secretary of the Society, George Prichard, outlined a campaign based on a two-pronged attack dedicated to maintaining stability: one, against unrespectable knowledge, and two, against unrespectable sex. The boundary between these was far from clear, and both were considered to be potentially disruptive. Prichard said that the Society's aims were to prevent blasphemous publications, obscene books and prints, disorderly houses, the profanation of the Lord's Day, and fortune-telling.[23] About fortune-telling, he suggested specifically that the Society disapproved of it because it encouraged a fatalistic belief in the outcome of predictions, leading to apathy and a 'lack of good conduct'. Fortune-tellers, he said, reduced their 'credulous dupes' to debt, and then incited them to the robbery of their employers. Further, those who fell under the sway of fortune-tellers were 'most commonly servant girls', who would allow themselves to be reduced to complete destitution, 'even to the disposal of their wearing-apparel'. This is almost certainly a reference to prostitution. There was a widespread belief that brothel owners used a technique of trapping innocent young girls into their service by issuing them with clothing to replace their own. If the girls ran away, they could then be arrested for theft of the clothes and brought back. This trick was referred to in the 1750s at court cases heard before Henry Fielding, and was cited again in the campaign led by W. T. Stead at the end of the nineteenth century.[24]

The fact that servant girls in particular were regarded as being at risk demonstrates that it was the middle-class home which was thought to be particularly vulnerable. Servant girls were thought to be extremely unreliable, and the impossibility of keeping them was a major preoccupation of respectable families, as demonstrated in many pages of advice in household manuals. One study has found that the median period of service for all domestic servants at the London Foundling Hospital between 1759 and 1772 was a mere 20 weeks, for scullery maids only 10 weeks.[25] Employers feared that servant girls 'alternated between service and prostitution', and were highly anxious about women servants' notoriety for frequently changing their place of employment.[26] They might bring the consequences of their misdemeanours into the employer's private sphere, a danger of which the rising urban middle classes were all too aware.[27] One of the questions frequently asked of Powell, so it was claimed at his trial, was 'Whether servant or lodger may be trusted?'[28] This question, incidentally, suggests that it was not only servants who were liable to fall prey to superstitious belief, since it was an enquiry put to the astrologer by employers.

It was suggested that fortune-tellers targeted servant girls. Those who sold fortune-telling door-to-door, like the match-woman advertising Joseph Powell's services, often began by offering 'petty articles' for sale. This was said to be a safety device: 'if the master or lady of a house opened the door, and they had nothing to offer, they would attract suspicion'.[29] It was the young female domestic they were really hoping to speak to, knowing that she was the member of the household most likely to be interested in what they had to offer, and hoping, perhaps, to hear some secrets about the neighbourhood in the course of idle gossip on the doorstep. This, in turn, would allow them to pretend great knowledge from a supernatural source when they spoke to a neighbouring maid, winning the foolish girl's trust.[30] Both sexual propriety and material security were at risk.

However, the Society believed that not only did fortune-tellers reduce their 'victims' to desperate measures which led to immorality, but that they were also more directly involved. When Thomas Barnes called on Powell at home, he was shown into a front room to wait because the astrologer 'was engaged with some females in the back room'. A letter in Powell's handwriting was produced in court in which he answered a lady's enquiry about her ability to have children by telling her that: '[Y]ou may have Children – but it must be by some other Person. I should like to be that one, to enjoy that pleasure with you. And I doubt not but what we should – make an increase.' If she wished to agree to this, the letter continued, Powell (who, according to the Society's account, already had three or four wives) would calculate her nativity free instead of for the usual five guineas. All this evidence seems to justify the widely held belief that the most frequent customers of fortune-tellers were young women anxious about love and ready to be taken advantage of. This might perhaps have been the reason that Blair made the focus of the letter he sent to Powell questions about the young Ann Mintridge, in order to lend credence to Barnes's enquiries.

The association between lack of knowledge (fortune-telling) and lack of morality (promiscuity and bigamy) was very strong in the Society's understanding of disorder. One of its most distinguished later members, Hannah More, made this point in several of her Cheap Repository Tracts, published to help educate and warn the poor of the dangers of ignorance and superstition. One of these was called *Tawny Rachel*, and was about a fortune-teller and her eventual prosecution and transportation. Told in the usual chatty and almost racy style of More's tracts, *Tawny Rachel* told the story of the wife of 'poaching Giles', who 'had a sort of genius at finding out every unlawful means to support a vagabond life'. Rachel travelled the country with a basket on her arm, pretending to sell laces, cabbage nets, ballads and history books, and to buy old rags and rabbit skins, but in reality this was only a pretence

> for getting admittance into farmers' kitchens, in order to tell fortunes. She was continually practising on the credulity of silly girls; and took advantage of their ignorance to cheat and deceive them. Many an innocent servant has she caused to be suspected of a robbery, while she herself, perhaps, was in league with the thief. Many a harmless maid has she brought to ruin by first contriving plots and events, and then pretending to foretell them.

Rachel was, of course, eventually brought to justice. She was finally captured and taken into custody just at the dramatic moment when she was about to wreak the most damage, 'dealing out some very wicked ballads to some children'. She was sentenced to be transported to Botany Bay, 'and a happy day it was for the county of Somerset, when such a nuisance was sent out of it'.[31] The message was clear: superstition was not simply some 'harmless' fun that an idle wife might dabble in, but a moral danger to the innocent.

A very specific danger which the Society felt was posed by Powell was his encouragement of the vice of gambling. Predicting the outcome of financial speculations was part and parcel of fortune-telling work. This might include law suits, wagers and cockfights as well as the lottery.[32] Powell, however, seems to have had a lucrative trade in advising which numbers to buy:

> One of the chief sources of profit to this impostor, seems to have been the sale of 'lucky numbers' in the lottery. It is really surprising to observe, what a vast variety of memorandums he had relating to this subject. The evils attendant on the state-lottery must be very great indeed, among the lower classes of mankind; and it is much to be wished that means could be soon devised of abolishing this species of legal gambling.

The evils which were thought to ensue from gambling were, according to the Society, those that resulted from 'the manner of obtaining that money which they were so induced to mis-spend'.[33] Gambling, in other words, led to theft. However, middle-class disapproval of gambling was much more complex than this simple equation offered by reformers. Lotteries, in particular, were a focus of concern. Unlike the state lottery, private lotteries were illegal, but they were known to be widespread. One variant was the gambling that took place on lottery insurance, a cheaper and more widespread form of betting: 'In return for, say, a shilling, a pound would be promised if a certain specified number turned up.'[34] It was said that those most likely to take part in lottery insurance were 'male and female domestic servants; indeed, it was computed in 1800 that, on an average, each servant in the metropolis spent, annually, as much as twenty-five shillings in this vile practice'.[35] Such behaviour, even if servants spent their own earnings, was widely regarded as bound to lead to ruin. The instability of servants was one of the nostrums of the age. Their 'roving disposition' was loudly condemned, as was their tendency to spend valuable free time in dissolute places like the coffee shop or pub, places where the 'Morocco men' working for lottery agents (so called because of the red morocco pocket-books in which they kept their records) might prey on them.[36] A typical cautionary tale was that of the servant of a 'lady of quality', who 'under the absurd infatuation of a dream, disposed of the savings of the last twenty years of his life in purchasing two lottery tickets, which, proving blanks, so preyed upon his mind that, after a few days, he put an end to his existence'.[37] This story illustrates not only the dreadful dangers of financial irresponsibility, which the middle classes feared, but also an attempt to find some rational explanation for behaviour that might otherwise seem incomprehensible. To risk all one's life savings on a gamble could only be explained by some act of unconsciousness, almost of madness. To attribute the motivation for this act to a dream was to remove it from normal, voluntary behaviour to the pathological.[38]

Another form of gambling, which grew enormously in popularity throughout the nineteenth century, was horse racing, a phenomenon which caused 'increasing alarm amongst both religious and secular observers'.[39] Even later, during the Depression of the 1930s, betting on the horses remained high on the list of working-class spending. Curiously, historians of these two different phases of gambling differ in their interpretation as to whether the practice signalled pessimism or optimism about the future. For David Vincent, writing about the nineteenth century, the regular outlay of small sums on races 'combined entertainment, mental exercise and the real prospect of at least the occasional intermission of the cycle of deprivation. It made sense, but the pleasure of the pursuit rested on an underlying pessimism.'[40] For both Ross McKibbin and Gary Cross, however, writing about the 1930s, 'Though the working-class attitude to time was probably fatalistic it was also optimistic.'[41] This apparent difference is surely only a matter of how long a period of time we are considering. Some sort of short-term optimism is necessary in order to purchase a lottery ticket or bet on a horse. However, the longer term pessimism of what McKibbin refers to as fatalism is implicit in the desire to look for luck to improve one's finances.

In fact, reformers disapproved as much of fatalism as they did of immoral behaviour. In contrast with middle-class, educated views of the future, which were supposedly based on a happy trust in the outcome of industry and thrift, many commentators expressed the belief that plebeian, or uneducated, ideas about the future were likely to be pessimistic and fearful.[42] Criticisms of almanac readership, for example, denounced astrology for perpetuating belief in predictions of gloom and disaster. Generations of readers followed the predictions of death regularly included in the pages of *Merlinus Liberatus*, where the astrologer-compiler, John Partridge, analysed birth charts and deduced the time of death of the client.[43] In his important study of religion in rural Lincolnshire, James Obelkevich describes the astrological practice of John Worsdale: 'His books contain scores of examples of his forecasts, most of them pessimistic.'[44] Even in non-astrological chapbook prophecies, such as those claiming to come from Joachim of Fiore, Mother Shipton, or the Sibylline oracles, the overall tone was one of warning and disaster to come. Of course, such a tradition stemmed from the biblical prophets, whose function was to remind an erring nation of the imminence of God's anger and punishment. Prophecy and prediction in popular culture fulfilled a warning role, and thus were chiefly devoted to painting a picture of a future filled, at least in the short term, with disaster and suffering.

Disapproval of the apparently pessimistic nature of popular belief was expressed in the Powell case, when the Society noted that the nativities seized in the astrologer's papers 'contain very dismal prophecies of what would take place in future years, and sometimes even pretending to foretel [sic] the death of different individuals'. It particularly disapproved of those instances when clients consulted him about the time of death of their parents, since 'unthinking or ill-instructed children might be tempted to act very improperly'. It went to the trouble of compiling a list of questions frequently asked of fortune-tellers, based on the papers in Powell's collection, the first of which was, 'Whether father or mother will die first, and when?', followed by 'Whether fortune or legacy will be left to the inquirer?' This was a peculiar variation on the theme of the doom-laden nature of prognostication, crediting it with a weakening of ties of family attachment. Again, in the Society's eyes, mistaken knowledge, or 'superstition', seemed to have dire consequences on an aspect of the private sphere which was vital to the security of public order.

Reformers also believed that a fatalistic attitude exacerbated the behaviour of the drunkard, the profligate, and the idle, by decreasing their hopes in the future and limiting their capacity for self-control. In 1817 George Prichard declared that the greatest danger of fortune-telling was the promulgation of a belief in fatalism, which was counter-productive to the benefits of hard work. One of the fundamental tenets on which the idea of progress depended was the importance of the development or improvement of the human character. As Martin J. Wiener puts it: 'Many movements to promote control over one's bodily habits flourished in the nineteenth century, from temperance, vegetarianism, and sexual purity to the "practical science" of phrenology. All were preoccupied with helping individuals aid their future well-being by developing control over their impulses and appetites.'[45] Self-control led to self-help which in turn led to self-improvement. This is demonstrated again in the Society's published account of Powell's trial, which could well have concluded with the magistrate's sentence, as the defendant received six months' detention with hard labour; but an even later exchange of words is recounted. As the prisoner is led away, he turns to say, 'I hope the Court will let me have the books that were taken from me.' Clearly Powell is not a reformed character, since the only possible use of the books of reference and ephemerides, which had been seized and handed over to the Society, would have been to continue his astrological practice. The Society suggests that his earlier apologetic response had been designed to persuade the magistrate not to send him for transportation, the most severe penalty he might have faced if the court had been unsympathetic. He is therefore told that the books will not be returned. The inclusion of this final detail is a clear acknowledgement of the role of reform. Powell is not simply being punished; he is being taught that his old ways of thinking will do him no good. The trial is not only about vagrancy, nor even solely about fraud, but about the close link between acceptable knowledge and self-improvement.

Nevertheless, social reformers' ideas about self-improvement did not always coincide with other people's. Despite the emphasis placed by the legislation and the Society on financial aspects of the fortune-teller–'victim' transaction, most of the questions about which Powell's clients consulted him were concerned with marriage, partners in love, and the bearing of children. Indeed, this was the only possible road to 'self-improvement' for many women. One recent study of early modern courtship and marriage stresses that not only was the choice of a marriage partner a crucial one, it was also one stage of life in which some women felt they could actually exercise some independence and agency:

> [T]he adolescent mythology of courtship foster[ed] a romantic ideal which exaggerated female authority and control. According to literary models, which circulated among plebeian women as well as elite female readers, courtship was the only female life-stage in which the usual relations of dominance and subjection between the two sexes were inverted. While the transient process lasted, women were supposedly 'on top' of the gender hierarchy.[46]

Amongst the 'ridiculous and unimportant questions' which the Society found in Powell's papers were the following:

> Whether the querist ought to marry, and whether husband and wife will agree?
>
> The state of life, personal qualities, and pecuniary circumstances of the person who offers to marry?
>
> How to — a male or a female child? [The Society deleted the word 'conceive'.]
>
> Is a woman pregnant or not; and if pregnant, whether with a male or female child?
>
> The fit time to buy cattle, hire a servant, visit any person, or make love to a woman.[47]

The seriousness with which enquirers might take the answers to such questions is demonstrated by another example from the life of Samuel Bamford, an early experience which might go some way to explaining why he remained so fascinated by the subject of magic in later life. In his autobiography he relates how, as a young man, he fell in love with, but was rejected by, a young girl, who, despite all his pleading, insisted that they should part. In response to his desperate requests for an explanation, she told him that she had 'given ear to the prophecies of an old fortune-telling woman', who had told her that

> 'it was not our fate to be united,' – 'that if the connexion was not broken off, one of us would die,' – 'that an evil star was in the table of our destiny,' – 'that, in fact, if the acquaintance was continued, I [Bamford] should prove false in the end.' 'And so,' added the distressed and almost terrified girl, 'what must be, must be;' – 'It is of no use striving against the decrees of Providence.' 'It is a great misfortune, but it might have been worse.'[48]

In vain Bamford tries to convince her that her belief in the fortune-teller's predictions is a delusion. The old woman, he is told, is always right, that 'whatever she ... foretold, it was useless to attempt to evade'. Even re-creating this bitter disappointment from the perspective of 30 or so years later, it does not seem to occur to Bamford that the story might have been his young woman's means of letting him down lightly. He feels thoroughly rejected, but never doubts the power of fortune-telling over the decision.

In the case of Joseph Powell, the questions he was asked about love and marriage no doubt betrayed a great deal about his clients' personal problems. The fact that he claimed to be able to advise on such private matters must surely have made his pronouncements seem impudent and improper to critics, who might well express dismay at 'the alarming and immoral consequences of fortune-telling'.

The fact that astrologers were driven into secretive practice in their own homes during the nineteenth century highlights the peculiarity of the use of the Vagrancy Act to pursue them. Those prosecuted were not vagrants, but this observation seems to have carried no weight. In 1851, during the trial of another astrologer under the same legislation, the defending solicitor tried to stress the difference between fortune-telling and astrology. His client, Francis Copestick, was a student of science, who, he said, was able to deduce from astronomical phenomena 'their influence over the future conduct of mankind'.[49] He was a settled householder, owning his own house in Bath, and paying rates and taxes. Despite this defence, Copestick was sentenced to a month's imprisonment with hard labour.

Powell does not even attempt to argue his own respectability, but takes great pains to show deference to the court and to accept its terminology. He speaks as if vagrancy and unemployment are the main issues. Yet somehow the Society and the magistrate detect the quiet strategies of resistance. Astrology was not usually regarded by its practitioners as a form of fortune-telling, and it is unlikely that Powell regarded himself as a rogue or a vagabond. In the handbill which brought about the prosecution, he described himself as a 'professor of sidereal science', a title which laid claim to an elite and ancient knowledge. There were certainly fortune-tellers who wandered the streets using cards or palmistry or coins to tell the future, but these were very different practices from those of calculating star charts and pronouncing on character. Powell's great mistake was in asking for his books back, demonstrating that he had not after all identified with the epistemologies of the court.

The elision between the different beliefs and practices of unorthodox belief under the broad rubric of fortune-telling or 'superstition' is illuminating for what it reveals about the tendency of dominant epistemology to subsume differences into a homogeneous 'othering' of oppositional knowledges, and points to the unhelpful nature of 'umbrella' terminology. The word 'magic' and its associated category 'the occult' are sometimes used in modern analyses as metahistorical labels which distract from specific beliefs. Even a recent and highly regarded monograph, the excellent study of late nineteenth-century spiritualism by Janet Oppenheim, *The Other World*, uses the term 'magic' in a strangely imprecise manner. 'By the nineteenth century,' she writes, 'magicians had become purely secular figures, entertainers whose function was to amuse ... but certainly not to heal ... predict the future, locate lost articles, or apprehend a criminal.'[50] This is an extraordinary conflation of the nineteenth-century phenomenon of conjuring, on the one hand, and healing, prediction, and astrology, on the other. The fact that Joseph Powell was caught up in legislation to control vagrancy and fortune-telling should not distract from the fact that he had been practising as an *astrologer* for 30 years.[51]

The modern urban environment did, without doubt, play an important role in the early nineteenth-century campaigns against fortune-telling and astrology, but it was in a more complex fashion than as the agent of rationalism.[52] Joseph Powell was sentenced to six months' imprisonment with hard labour not simply because he held 'superstitious' beliefs, but because material changes, in the form of the rapid growth of the population of London and attendant fears of public immorality and disorder, made his prosecution seem useful to an interest group which associated superstition with ignorance and crime. Further arrests of fortune-tellers and astrologers in their own homes, as well as the burgeoning sales of do-it-yourself fortune-telling guides, examined in the following chapter, demonstrate that 'magic' continued to withdraw into the private sphere.[53] In fact, its withdrawal signals an expansion of the public sphere, with what had once been considered private business now subject to prosecution. The privatisation of religious belief has been a phenomenon identified by sociologists and historians of religion as taking place later in the nineteenth century.[54] This unorthodox and early manifestation of it, as astrologers withdrew from the public eye, underscores the message we have already seen advocated by calendar publishers about society's increasing emphasis on the individual's responsbility for his or her own future: a rational and well-planned private life would contribute to the well-being of the wider social body.

3
The Interpretation of Dreams

The concept of dreaming as a sublimated expression of desire is now well known, following Freud, but at the beginning of the nineteenth century the interpretation of dreams related to more than just unfulfilled longing. It was an important aspect of the fortune-teller's work, a window onto the future. Amongst the papers of Joseph Powell were several notes recounting 'the dreams of different individuals, especially concerning lottery tickets'. Powell's clients wanted to know if their dreams about winning numbers were omens of future success. Such interpretation was not a straightforward matter, since long tradition suggested that dreams went by 'contraries', that is, that the opposite of what you dreamt was to be expected. Yet the rules were not simple, and a skilled 'oneirocritic', an interpreter of dreams, was as respected and valued in the nineteenth century as was Joseph in the days of Pharaoh. This, too, was an area of plebeian culture that was subjected to reform efforts, but it was an area where change was particularly difficult to achieve. Prosecution was not a useful tool, since dreams and their interpretation were amongst the most secret of consultations. Instead, the middle-class press advocated education and the advance of enlightened attitudes. This chapter examines the role of dreams in the representation of the future, and outlines how dreaming's claims on prophecy were undermined by the advance of rational knowledge.

The Status of Dreams

Dreams were a 'key theme of nineteenth-century European thought', preoccupying writers, philosophers, and psychiatrists.[1] Tony James has examined the important role of sleep and its associated phenomena in nineteenth-century discussions about the nature of human consciousness. Waking and sleeping were generally

Figure 7: *The Ladies' New Dreamer* (1890)
The title page of a typical dream book. The pictures around the woman's
sleeping form show the subjects which the compiler believed to be her
most likely concerns: love, marriage, and prosperity. As for the meaning of
these images: one dream book advises that 'If a young woman dreams of a
dagger she will rule her sweetheart in any manner she chooses.' Another
says that to see a serpent rising up out of the ground shows that the
dreamer has a female enemy.

represented as two distinct states: 'The first was valued because linked with consciousness, reason, and will; the second not, because divorced from these "higher functions".'[2] The exercise of will and active judgement, valued more than involuntary action, were crucial in the conceptualisation of maturity and health. The dominant scientific opinion throughout the nineteenth century was that dreams were an indication of disturbance and ill health, and that they did not occur in the deep, normal sleep of a healthy person. *Chambers's Journal* stated categorically in 1876: 'No one dreams when he is sound asleep. Dreams take place only during an imperfect or perturbed sleep.'[3] Carl Binz described dreams as 'somatic processes which are in every case useless and in many cases positively pathological'.[4] Since dreaming was associated with lack of control, both it and daydreaming, its related activity, were on the very borders of respectability. Warnings were issued about the dangers of indulging in reverie, 'a dangerous mental activity that often led to "an incurable habit of inattention"'.[5] In Enlightenment thought an inability to distinguish dreams from reality had become a mark of subhuman existence, a feature of animals.[6] The nature of humanity itself, therefore, might be defined as the capacity to differentiate dreams from true sensations, and reverie might endanger the boundary between the two. Terry Castle has suggested that several scientific studies imply a link between reverie and masturbation, both with awful effects on mental health.[7] In an 1865 study, *The Literature and Curiosities of Dreams*, Frank Seafield included a chapter entitled 'Analogies of Dreaming and Insanity', and Freud devoted the final section of the first chapter of *The Interpretation of Dreams* to 'The Relations between Dreams and Mental Diseases'.

Yet scientists were well aware of the long tradition that dreams could be about the future. One 'scientific' explanation was that dreams could at times furnish the answer to problems by the memory working involuntarily on the situation and producing a clarity in the reworked structure of events. In 1802 Georges Cabanis related a story about Benjamin Franklin in which Franklin believed that dreams had 'informed him about the outcome of affairs which were preoccupying him': '[H]e had not been able to avoid superstitious ideas about these internal warnings, not realizing that his own capacities of prudence and sagacity were still functioning during sleep.'[8] In keeping with this dominant scientific paradigm, only those writers who were prepared to be associated with lack of control, challenging the norms of middle-class masculinity of their day (Balzac and the occult, Baudelaire or Coleridge and hashish), dealt seriously with dreams as omens.[9] Mystical and literary thought never lost its fascination with altered states; for example, Emanuel Swedenborg, the influential Swedish mystic and visionary, had kept a *Journal of Dreams*, which he wrote in 1743 and 1744, and this was finally published in 1859. Indeed, no one doubted that there had been great men in the past who had experienced prophetic dreaming, but just as prophecy was an activity which became relegated to history, so prophetic dreaming's greatest days were clearly over. When the contributor to *Blackwood's* in August 1840 discussed the topic under the heading 'A few passages concerning omens, dreams, etc.', he related stories of men who had premonitory dreams, but addressed his thoughts to an imaginary Eusebius, clearly relegating the subject to a classical past.[10] The interpretation of dreams was no longer a respected profession. Seafield wrote: 'Oneiro-criticism is at present in the sere and yellow leaf of its fortunes. It sprang up to meet us like a god; it retires from us with the hang-dog expression of a rebuked costermonger.'[11]

Dreaming, then, was predominantly associated with mental or physical disturbance. As such, in the latter half of the century, it was a topic often linked to women's ill health. Women's association with hysteria and altered states of consciousness was depicted by medical opinion as connected with menstruation and physiological weakness, and disturbed sleep which included dreams was part of this pattern of neurotic disorder: 'The accumulation of nervous energy which has had nothing to do during the day, makes them feel every night, when they go to bed, as if they were going mad.'[12]

The interpretation of dreams was likewise frequently linked to women, whose own dreaming clearly fitted them best for the task of prognostication. In one of Samuel Bamford's stories of his younger days in Lancashire, he recounts the tragic events surrounding the daughter of a prosperous farmer, a beautiful young woman with ambitions to marry higher than the station of the young dragoon who wooed her. Later, either from 'some reproaches of conscience, or from some "dream or vision," or some "apparition," or "love spell"', she suddenly decided to travel to Scotland, where the soldier was then quartered. As her horse entered the town, a funeral passed by, and, on being told that it was her own rejected lover who lay in the coffin, she collapsed, and never fully recovered. When Bamford knew her, in the early nineteenth century, she had become an old woman, wandering from place to place, exciting both pity and dread in those who observed her, but known to have a 'remarkable gift of prayer' and to be an interpreter of dreams. Her loss of 'mind' and her dishevelled appearance, 'face half muffled ... in an old brown cloak ... with sundry rags, bags and pockets', served to increase the respect which she was accorded, as a living example of the way in which madness could open the mind to a higher power.

Even in middle-class accounts of prophetic dreaming, it was most frequently women who were the dreamers. In 1833, in just one example of many in the periodical press, *Chambers's Journal* carried a story of a young man's life being saved by the dreams of his aunt, who saw the fishing boat he was due to go out in sinking. She had this dream three times in one night, and in the morning entreated him not to go. He followed her advice, and learned later that the boat had sunk, and all in it had been drowned.[13]

It is strange to find stories of prophetic dreaming appearing in the periodical press at a time when prosecutions against fortune-tellers were gathering momentum. One reason was that the highly personal nature of predictive dreaming removed it a little from the sphere of dangerous knowledge. Moreover, the stories in the periodical press, like the *Chambers's Journal* article, were predominantly about middle-class dreamers whose premonitions saved their loved ones from danger. The dreams recorded from more plebeian dreamers were generally more political, and more subject to disapproval. For example, in the early penal settlement of Australia, the dreams of an old woman, a convict from Scotland, caused unrest. There was a rumour that several French warships were on their way to destroy the settlement and to release the convicts. This was considered perfectly feasible, since England was at war with revolutionary France. Governor Phillip discovered that the story had its origins in an old woman's prophecy, but when he spoke to her personally, she protested that she had only been recounting a dream, and had never intended that it would be taken for a prediction. She was a 'harmless old creature', a native of Scotland, who was anxious to clear her good name. All sorts of extraordinary prophecies had been attributed to her which she absolutely denied – they had been only dreams, she said. The governor listened to her distressed explanations, said that he understood, and told her that she was not to worry. From his lofty position of educated knowledge he was able to dismiss the old woman's worries that the unsettling effect her dream had had on those around her would be seen as dangerous. No further action would be taken, he promised, though its possibility hovers as an unspoken threat in the account given by David Collins, the solicitor-general of the colony, and the very fact of a visit from the governor may well have seemed in itself a threatening gesture.[14] A significant feature of this episode is the way in which Collins links the power of the old woman's dreams to those who particularly responded by believing in them: the Irish convicts. While saying very little about this aspect of the incident, Collins demonstrates the assumption that superstitious beliefs could most readily be found amongst the rural poor, a section of the population which, particularly in its Irish context, might be most dangerous to 'civilised' values. Similarly, in England, the radical dreams of Joanna Southcott and Richard Brothers, the one dreaming of war and pestilence, the other of rivers running with blood, were responded to as threats. Southcott was shunned by the church and government; Brothers was locked away in prison.[15]

Accounts of middle-class dreams, in contrast, remain resolutely about the personal rather than the political. For example, William Howitt, the author and spiritualist, had a dream while on the journey out to Melbourne of what his brother's house would look like. He told the dream to fellow passengers and to his sons. When he arrived, the house was exactly as he had dreamt, looking out onto the woods he had seen: 'When I look on it, I seem to be looking into my dream.'[16] In 1841 Georgiana McCrae, best remembered for her work as an artist, noted in her diary:

> Mrs. Thomas ... came to see me and to tell me of a remarkable dream and its fulfilment. At breakfast she told David she had dreamt that two Scots cousins of his had arrived. After the doctor had gone out, a Mr. Lawlor called and announced himself as a Scots cousin – husband of Ann Thomas, of Edinburgh.[17]

There were few more upright and rigid representatives of Christian respectability than J. D. Lang, the first Presbyterian minister in Australia. Yet even he implied in his memoirs that he had been visited with a dream of prophetic import. On the night that his father was lost at sea, Lang dreamed he saw him standing on deck and the ship going down. This may be interpreted by the rational mind as nothing more than worry, and Lang's father did indeed drown, but Lang implies that the dream had something special about it, if only for the fact that he very seldom dreamed.[18]

These respectable accounts, apart from relating to private, family concerns, have one other important characteristic in common: in each case the dreamer finds it transparently easy to draw connections between the vision and a later event. There is little metaphorical or symbolic imagery requiring interpretation. For more complex dreams, the meaning of which seemed unclear or ambiguous, and where the 'event' to follow had not yet made itself clear, there were two possible courses of action: to consult a wise or cunning man or woman, or fortune-teller; or to attempt an interpretation using a chapbook as a guide. Some aspects of consulting a fortune-teller have been examined in Chapter 2, where the dangers of confiding secrets to someone who might be prosecuted and imprisoned were highlighted. Here I will concentrate on the other recourse available, especially to those who could not afford to consult an 'expert' in prediction: fortune-telling literature.

Books of Fate and Dream Books

In the middle of the nineteenth century, the radical James Guest described how street pedlars sold a wide array of different types of books. Hawkers, he wrote, would sell

> at what they could get, at prices varying from 2d. to 6d., sometimes a good supper and leave to sleep in the barn or outhouse, often when they could get the blind side of the old dame or the young one with their Pamphlets, Books of Dreams, fortune telling, Nixon's prophecies, books of fate, ballads, etc.[19]

Guest's list reveals several facets of popular fortune-telling literature which deserve closer examination: its gendered nature ('the old dame or the young one'), its claims to antiquity (the seventeenth-century 'Cheshire prophet', Robert Nixon), and its cheapness, which made it the reading matter of the masses, either newly or barely literate.

Growing literacy did not detract from a belief in the potency of predictive rituals. In the 1840s a visitor to Abel Heywood's successful publishing business in Manchester noted that amongst the 'literary chaos' of Heywood's shop were, along with 'masses of penny novels', sectarian pamphlets and democratic essays, a host of dream books.[20] In 1891 Charles Leland, the noted antiquarian who was well known for his studies of gypsy society, wrote that there were 'many millions more of believers in such small sorcery now in Great Britain than there were centuries ago', and lay responsibility for this development with the spread of print culture:

> There is to be found in almost every cheap book, or 'penny dreadful' and newspaper shop in Great Britain and America, for sale at a very low price a Book of Fate – or something equivalent to it, for the name of these works is legion – and one publisher advertises that he has nearly thirty of them, or at least such books with different titles. In my copy there are twenty-five pages of incantations, charms, and spells, every one of them every whit as superstitious as any of the gypsy ceremonies set forth in this volume. I am convinced, from much inquiry, that next to the Bible and the Almanac there is no *one* book which is so much disseminated among the million as the fortune-teller, in some form or other.[21]

David Vincent has described the impetus that literacy gave to the mass production of 'penny do-it-yourself fortune tellers', which 'saved the newly literate from the embarrassment and expense of laying their difficulties before the familiar but strange figures of the wise men and women'.[22] He might also have added that it saved them from contact with the local constable.

Any chapbook that claimed to predict the future tended to be subsumed under the broad rubric used by Leland, the 'book of fate'. However, such chapbooks did fall into certain distinguishable types. Some concentrated on active intervention, remedies and charms, guides to influencing the future: these might include what they called 'recipes'. Others dealt more with different ways of reading the future, from signs in the natural world to the casting of dice, as well as lists of fortunate and unfortunate days of the year. Many included brief sections on physiognomy, particularly the reading of moles ('If a man hath a mole athwart his nose, he will be a traveller').[23] Dream books, or dreamers, did exactly what their name suggested. They provided an A to Z of meanings of dreams; for example, the appearance of comets in a dream 'is ominous of war, plague, famine, and death', to dream of a cat signifies that you will soon catch a thief, beer is a portent of an accident, and crows flying in cloudy weather show coming loss and misery.[24] Although some writers expressed the belief that these meanings were derived from Artemidorus of Ephesus, whose work on dreams in the second century was thought to be the earliest surviving example of the genre,[25] dream books no doubt fed off one another, just as all street literature did. It was possible for the same material to be recycled in several different forms. Attempts to induce certain dreams might, for example, be included in the 'recipes' of the fortune-telling book.

Many dream books demonstrated a familiarity with the burgeoning discussion of dreams from a physiological standpoint. Those chapbook writers who aspired to respectability made sure their readers knew the extent of their knowledge of current discussion, if only through a preface. The compiler of *The True Fortune Teller* of 1850, for example, printed in Edinburgh, issued the following disclaimer:

> TO THE READER
> The foregoing pages are published principally to show the super-stitions which engrossed the mind of the population of Scotland during a past age, and which are happily disappearing before the progess of an enlightened civilisation. It is hoped, therefore, that the reader will not attach the slightest importance to the solutions of the dreams as rendered above, as dreams are generally the result of a disordered stomach, or an excited imagination.[26]

Some compilers achieved a similar distancing effect by publishing titles that they claimed had come from abroad, noting that the books were translated from the original Greek. The *High German Fortune-Teller* and *Napoleon's Book of Fate* were both said to be translated from German. Others used dubious pedigrees of ancient lineage, which conferred some kind of authority, but also served to remove respon-sibility for publication.

Any form of predicting the future, whether prophecy or fortune-telling, claimed to carry the authority of long-established custom. The belief that these chapbooks tapped into an old tradition is demonstrated by the appearance on James Guest's list of the name of Robert Nixon, supposed Cheshire 'prophet' of the seventeenth century. The astrologer Raphael, otherwise known as Robert Cross Smith, claimed that the dream book he published (*The Royal Book of Dreams*) was derived from an ancient manuscript which he had simply stumbled across while out on a country walk. In the summer of 182– [sic] he had come across a broken-down Somersetshire courthouse, and was there shown a 'curious manuscript, which was buried in the earth for several centuries, containing one thousand

Figure 8: *Napoleon Bonaparte's Book of Fate* (1850)
Perhaps the most popular fortune-telling book of all; there were many versions, most of which claimed to be translations of a book 'written in German nearly 500 years ago', 'a cabinet of curiosities, and valuable secrets', which had been seized, so it was claimed, from the belongings of Napoleon Bonaparte after his defeat at the Battle of Leipzig.

and twenty-four oracles or answers to dreams ... whereby any person of ordinary capacity may discover these secrets of fate, which the universal fiat of all nations ... has acknowledged to be portended by dreams and nocturnal visions'. Similarly, in 1850, the publisher of *The Dreamer's Oracle* claimed to have found the original of his dream book 'in the Ark of a Late Celebrated Wizard'. Throughout the nineteenth century repeated issues of *Napoleon's Book of Fate*, perhaps the most popular fortune-telling book of all, claimed to be a translation of a book 'written ... nearly 500 years ago', 'a cabinet of curiosities, and valuable secrets', which had been seized, so it was claimed, from the belongings of Napoleon Bonaparte following his defeat at the Battle of Leipzig, after he had used its wisdom to achieve power.

Claiming an ancient pedigree meant that this literature could always pose as a form of popular antiquarianism, that branch of respectable scholarly knowledge which gathered examples of the beliefs and practices of 'the common people'.[27] In fact, such claims might have been intended to be seen as spurious: the passing off of contemporary works of literature as ancient was practised openly by some of the most successful writers. This was an issue in controversies surrounding the late eighteenth-century forgeries of James Macpherson (supposed translations of the work of Ossian, a third-century Scottish poet) and Thomas Chatterton (translations of a fake mediaeval monk, Thomas Rowley). There were, though, some who regarded the ability to fake a particular literary heritage as a skill in itself. Horace Walpole, the distinguished politician and author, had represented *The Castle of Otranto* (1764) as a sixteenth-century manuscript recently discovered. The lineage of dream books, therefore, whether taken as genuine or ironic (and, of course, compilers may have intended both, depending on the sophistication of the reader) helped to give a certain tone to the genre. This was helped by the fact that dreaming was perhaps the form of superstition that aroused the least disapproval, unlike palmistry and astrology, which were specifically forbidden by law. When Raphael claimed to have discovered his ancient fortune-telling manuscript in Somerset, he wrote that most of it had faded with time and was unreadable, but that he was delighted about this, since some of what fortune-telling taught (he mentions necromancy in particular) could be harmful. Only the most useful part, that on night visions, 'a helpful and proper area of study', remained intact and perfectly legible. This was, indeed, a very convenient situation for someone who, as his biographer writes, longed for respectability.[28]

Women and Superstition

> Whatever women may be, I thought that men, in the nineteenth century, were above superstition. (Marian Halcombe in Wilkie Collins, *The Woman in White*, 1860–61)

The role of women in fortune-telling literature was crucial. The compiler of the *True Fortune Teller* (1850) made clear that he could not be held responsible for the superstitions in his book, since he had found all the material in a cave in which a 'gypsey', old Mrs Bridget, or Mother Bridget, made her home. Mother Bridget was one frequently attributed source, Mother Bunch another, and Mother Shipton a third. These apocryphal sources were doubtless intended to attract readers in several ways: their age lent authority, their names suggested wisdom, perhaps through 'gypsy' connections, and their gender assured readers that their interests and needs were understood. As late as 1899 one London bookseller commented, 'I sell a most surprising number of publications of the prophetic almanac and dream book class ... The greater number of my customers in this way are workgirls and domestic servants, or young married women of the "small villa" class.'[29]

These books, then, provide a window onto women's culture of the nineteenth century. Although the contemporary suggestion that a belief in the visionary power of dreams might well be the preserve of women because of the link to medical constructions of physiological disturbance, the connection between women and 'superstitious' belief generally was more complex. As scorn became the dominant way in which the educated elite regarded fortune-telling and magic, it was increasingly accepted that such belief was due to a lack of education, and women of the lower classes were regarded as being amongst the least educated group in society. Certainly until the very end of the nineteenth century women were more likely than men to be illiterate.[30] When reformers campaigned against superstition, they frequently referred to the need to protect women from their own gullibility. Seafield wrote that dream interpretation 'is now an instrument by which a chap-book pedlar may best ascertain what is the smallest number of lies which Cinderella will insist on in return for her penny, without considering herself cheated?[31] Most of Joseph Powell's clients were women, and many of their questions were about dreams relating to love, marriage, and children. When the member of the Society for the Suppression of Vice who set out to entrap Powell wrote a letter asking him for an astrological interpretation, he centred it around questions about a young maidservant. When would she marry? Would she have children? Such questions, and from such an innocent source, were, we are led to believe, commonly asked of Powell. In *The Woman in White*, it is the low-born Anne Catherick, illegitimate child of a maidservant, who tries to warn Laura Fairlie by relating a prophetic dream: 'Do you believe in dreams? I hope, for your own sake, that you do.'

Even the compilers of books of fate regarded their publications as suited to the lower orders. The 1750 *Dreams and Moles* was designed, its title page proclaimed, to appeal to 'the very meanest Capacities'. One late nineteenth-century commentator, writing in 1890, made the point that different levels of society resorted to different forms of fortune-telling:

> The popularity of the Gipsies throughout the 18th century showed no signs of waning ... The amusement was very popular among a certain class: but that class was of a considerably higher rank than the people who indulge in the 'Dream Books' and such like of the present day.[32]

It is rare to find a contemporary distinguishing between different types of fortune-telling, but here, perhaps unexpectedly, we have the opinion that the published version of fortune-telling, the printed page, is more popular with the least educated. No doubt part of the explanation lies in the whole experience of visiting the gypsies, which would be exciting, novel, and even dangerous, a journey of social exploration to rival Henry Mayhew's journalistic travels through the streets of London.[33]

The contents of the dream books themselves confirm that women were thought to be the most likely readers. *Dreams and Moles* contains advice on 'How to restore a lost Maidenhead, or solder a crackt one' (a recipe using myrtle berries).[34] 'Charms for Dreaming' advised readers how to obtain a dream which would contain an image of their future marriage partner. In *Mother Shipton's Fortune Teller* of 1861, for example, there is 'A Charm for Dreaming' which advises the reader how to be sure of dreaming of the future by appealing to 'Luna every womans' [sic] friend'. *The Dreamer's True Friend* of 1861 includes 'how to choose a husband by the colour of his hair'. In *The Dreamer's Sure Guide; or the interpretation of dreams faithfully revealed* of 1830 a pull-out illustration shows a woman lying on a couch dreaming, with pictures of her dreams all around her. *The Dreamer's Oracle* also has a large illustration showing a woman dreaming of a handsome man.[35] The seeds of a wish-fulfilment interpretation of dreams are clearly here, with the difference lying in what these images actually represent. Are they visions of the future, appearing in dreams as prophetic insights? Or are they spectral imaginings, lurking in the unconscious?[36]

It may be that some women's lack of education predisposed them to irrational belief, but the existence of highly educated readers of dream books problematises such a simple explanation. Hesba Stretton, chief writer of tract fiction for the Religious Tract Society, was known to have slept with a dream book under her pillow.[37] And whereas the thrust of the legislation against fortune-telling was directed at vagrants, implying that the crime was most likely to be committed by the likes of the vagabond Tawny Rachel, whose exploits as 'a famous interpreter of dreams' were described by Hannah More, records show that most of those prosecuted were settled, even respectable practitioners. It is impossible, of course, to evaluate to what extent the purchasers of these little books believed or acted on their contents, and some may even have bought them simply for a laugh, as no doubt do some present-day purchasers of *Old Moore's Almanac*. Amongst Pepys's collection of chapbooks is the fortune-telling book *Mother Bunch's Closet*,[38] which claims its author as 'your loving friend, poor Tom', one of that large family of 'poor' authors who signalled, for those who understood, that here was a work of satire.[39]

Throughout the eighteenth and nineteenth centuries, as literacy and education advanced, the content of chapbooks did not greatly change. They were still about courtship, marriage, and the chances of good fortune. However, whereas the earlier versions had occasionally addressed a male reader, by the end of the nineteenth century the genre was targeted entirely at women. Marriage and children remained by far the most common subjects. 'A virgin dreaming she has put on new garments, shews an alteration in her condition by way of marriage.' 'If a woman dreams she is with child, it shews sorrow and sadness.' Margaret Spufford has related the popularity of courtship books in the early modern period to the late age of marriages, and other historians have noted that courtship 'comprised a major life-stage for the majority of the female populace, as women were left to their own resources [i.e. outside the family home in which they grew up] for a lengthy period between puberty and marriage at about age 25 or 26'.[40] In the nineteenth century fears of spinsterhood continued to be an issue for many women. The 1851 census showed that women outnumbered men, and this contributed to widespread discussion about the problem of 'surplus' women.[41] Small wonder that the question of marriage should figure so prominently in women's chapbook literature.

However, one very practical reason for women's interest in these books may have been their cultural role as guardians of the calendar. Throughout the eighteenth and into the early nineteenth century, women were the most likely purchasers of almanacs, perhaps to do in part with the timing of menstrual cycles and pregnancy.[42] In 1911 the London publisher Frank Smythson produced *A Mere Man's Calendar*, with the implication that most calendars were actually intended for women. Woman's 'natural' domain was increasingly defined as the home, with household management a major part of this role, and the calendar fulfilled an important function in domestic planning. The management of time was an important priority in the domestic sphere: 'The authors of housekeeping manuals ... made a well-regulated and orderly home the end of women's efforts. In order to create such a household, authors gave women advice about how to keep and to organise accounts; about how to arrange their day by the hour.'[43] The compiler of the first edition of that most rational of all calendars, *Whitaker's Almanac*, described it as 'essentially a Household Book' – this despite its enormous lists of statistics and government data. Although, of course, those who published household advice would never have countenanced a 'superstitious' use of calendars, nevertheless the association of calendar and unorthodox belief was a strong one, and may have found its way into many a 'respectable' home by default. The lists of lucky and unlucky days in fortune-telling books might have fallen into the role of planning in some homes, pointing to an area of women's influence, perhaps, over such matters as the timing of journeys.

Men had very practical duties to perform to guard against disaster in the climate of uncertainty to which commercial and professional interests made their families vulnerable, but women had very little role save in the emotional sphere of support. Fortune-telling, or loosely associated practices such as consulting 'lucky' days in the calendar, might be seen in this context. While many respectable women probably never considered 'fortune-telling' as a possibility, the boundary between acceptable and unacceptable advice and knowledge could be blurred. The distinguished historian of French popular literature, Henri Jean Martin, has categorised fortune-telling advice as a 'practical remedy'. Astrological delineation of character, including the propensity of certain astrological types to particular illnesses, he writes, often occurred in the middle of cookery recipes and even of prayers. Martin describes the contents of the French *Calendrier des bergers*, whose 'contents varied from commonsense advice to white (and even black) magic, the latter quite common in the early nineteenth century'.[44] In fact, he includes such astrological advice in a discussion of 'elementary science and how-to-do-it manuals'.

Condemnation of Dream Books

Dream books and fortune-telling books were disapproved of by all those who sought to bring education and improvement to the labouring classes. Both middle-class and artisan commentators believed that such literature locked readers into the hopelessness of a predetermined future. For some who disapproved, these books were also against religion. In one of Hannah More's tracts she warned 'all you young men and maidens' that 'God never reveals to weak and wicked women those secret designs of his providence', so that 'to consult these false oracles is not only foolish but sinful'. She listed the kinds of villains to take guard against as: *'cheats, imposters, cunning women, fortune tellers, conjurers, and interpreters of dreams'.*[45]

In terms of content, though, dream books were largely conservative. Wilkie Collins wrote of 'the helpless discomfort familiar to us all in sleep, when we recognize yet cannot reconcile the anomalies and contradictions of a dream', but dream books offered the calming reassurance that there was order and meaning in the apparent confusion.[46] At times they even appeared to recommend a view of society that was indistinguishable from that of the middle-class critics who deplored such literature. At least one advised its readers to abstain from beer. They could help to pacify unruly imaginations even when dreams made the fixed nature of gender seem questionable. *Dreams and Moles*, for example, described what it would mean if a woman dreamt that she was a man:

> When a woman dreams she is a man, and is not married, she will have a husband; or if she's without children she'll have a son, ... and to a maid-servant, much incumberance; 'tis verre fortunate to a harlot, because she will forsake her evil ways.[47]

A woman's dreams of manhood, then, signify the coming expression of her fulfilment. To a prostitute this is a very good sign – the man, no doubt, will rescue her. The maidservant, it is suggested, might have cause to regret her dream, since it heralds an unwanted pregnancy.

Even more potentially threatening than gender-subversive dreaming, however, was the question of sexuality. *Dreams and Moles* relates what it means for 'a barren woman' to dream that 'she prostitutes herself with her own sex'. In this case, the interpretation is again decidedly conservative: it simply shows that she will have a child. However, a 'fruitful' woman having the same dream would have 'much pain in bearing her children'. In an example typical of the way in which chapbook material was repeated across titles and across many years, this same interpretation occurs in the *New Infallible Fortune Teller* of 1818, but with slightly less confronting language: 'For a barren woman to dream she embraces one of her own sex, denotes that in time she will have children; but to a fruitful woman it denotes pain and sorrow in child bearing.'[48] Such interpretations position the meaning of dreams outside the workings of the mind, implying an external set of forces that will decide destiny. Yet in many nineteenth-century dreamers, messages of individual responsibility and the benefits of hard work can also be found, alongside reassurances about external determination. In this, the role of the dream book is ambivalent and complex, like much popular culture, seeming both to challenge and to validate the existing order. Several dreamers signal on their title pages that they contain the interpretation of dreams 'by the most ancient as well as the most modern Rules of Philosophy'. Popular literature was, after all, a commodity, and as such it was profitable to leave it open to as many interpretations as possible.

There is more to the unrespectability of dream books, however, than reformers' disapproval of superstitious beliefs or of fatalism. Dreaming is a notoriously atemporal activity, and any belief in the predictive capacity of dreams is a challenge to the fixed regularity of time. Many fortune-telling books, dreamers amongst them, included lists of fortunate and unfortunate days in the year to come, implying that the passage of time was not equal: the quality of days differed, even if the number of hours did not. A typical list of perilous days might say, for example, that 'on these days if a man or woman let blood, they shall die within twenty-one days following; and whosoever falleth sick on any one of these days shall certainly die; ... Also he that marrieth a wife on any of these days, they shall either be quickly parted, or else live together with sorrow and discontent.'[49] As many studies of popular belief have shown, certain days, especially those connected with the ecclesiastical year, were imbued with special significance;[50] and this was recognised in dream books. *The Golden Cabinet or the Compleat Fortune-Teller* of 1795 instructed the hopeful dreamer:

> On St. Valentine's Day, take two bay leaves, sprinkle them with rose water, the evening of this day lay them across your pillow. When you go to bed, putting on a clean shift turned wrong side outwards, and laying down, say these words softly to yourself.
>
> > Good Valentine be kind to me,
> > In dreams let me my true Love see.
>
> So crossing your legs, go to sleep as fast as you can, and you will see in your dream the party you are to wed come to your bedside, and offer you all the modest kindness imaginable.

Many of the charms and incantations designed to reveal a forthcoming husband depended for their efficacy on being carried out on particular days of the year. Rules were published by which interpretation of the character of children might be made according to what days of the calendar they were born on, referring to the age of the moon. Certain days were good for travelling on, others not.

These chapbooks were still being published, and being claimed as the most popular of street literature, at a time when the push for universal time was taking place at the level of government and international conferences. This is a very different economic context from that of chapbooks. The deterministic nature of the fortune-telling world-view was anathema to a society increasingly dependent on individual ambition and enterprise. Herbert Leventhal has written

BIRTH OF CHILDREN, AND OTHER EVENTS.

With respect to the Moon's age, and Day of the Week.

To be born on the first day of the new moon, portends their life shall be pleasant, with an increase of riches.

A child born on the second day will grow apace, and be inclined to lust.

A child born on the third day will be short lived.

The fourth day is bad. Persons falling sick on this day rarely recover.

The fifth day is favourable to begin a good work, and dreams will be tolerably successful.

The child born on the sixth day will not live long.

Do not tell your dreams on the seventh day, for much depends on concealing them.

The eighth day the dream will come to pass, and is a very prosperous day.

The ninth day differs very little from the former.

The tenth day is likely to be fatal ; those who fall sick will very rarely recover.

The eleventh day is a day to be married, or commence a journey. A child born on this day will be very healthy.

On the twelfth day, the child born will meet every affection, but be of a bad temper. It is a very unlucky day.

A child born on the thirteenth day will be unfortunate. Persons imprisoned this day will soon have their liberty.

A child born on the fourteenth day will die as a traitor.

The fifteenth day is very unfortunate.

The child born on the sixteenth day will be unfortunate.

A child born on the seventeenth day will be foolish.

The eighteenth day is very fortunate.

A nativity on the nineteenth day is fortunate.

Your dreams portend good on the twentieth day of the moon.

An unhappy fatality attends a child born on twenty-first day.

On the twenty-second day the child will be very fortunate.

On the twenty-third day the child that is born will be of an ungovernable temper.

On the twenty-fourth day, the child born will achieve many heroic actions, and will be much admired for them.

The child born on the twenty-fifth day will be very wicked.

On the twenty-sixth day, the child born will be very amiable.

A child born on the twenty-seventh day, will have every engaging quality.

On the twenty-eighth day, the child that is born shall be the delight of his parents, but have much sickness.

The child born on a Sunday shall be of a long life.

On Monday. Weak, of an effeminate temper.

On Tuesday. The person born will be very passionate.

On Wednesday. Shall be given to learning, and profit thereby.

On Thursday. He shall arrive at great honour and riches.

On Friday. He shall be of a strong constitution.

On Saturday. This is a bad day, though the child may come to good : they are of a dogged disposition.

Figure 9: 'Birth of Children, With respect to the Moon's age, and Day of the Week' *The Circle of Fate* (1850)
This list illustrates one reason for the popular demand for lunar information to be continued in calendars, when all other planetary detail was dropped. The moon was thought to have a wide-reaching terrestrial influence – on the weather, fertility, the growth of plants, female menstruation, health, and, as shown here, on the character of children. Eclipses, new, and full moons were particularly important.

about the way in which the practice of drawing up astrological charts to advise ships' captains when to set sail seems to have disappeared by the nineteenth century.[51] In the modern world setting out on a journey could not wait for lunar aspects or a propitious date.

As well as an irrational attitude to time and planning, fortune-telling chapbooks seemed to promote the kind of secrecy which contributed to the suspicion felt by the opponents of magic. The Society for the Suppression of Vice denounced '[T]he secret manner in which these astrological deceivers carry on their traffic.'[52] Yet books of fate were suited to private reading and encouraged rituals that centred around the bedchamber. The future might appear in a dream, and it was necessary to go to bed in the right frame of mind to receive the prophetic vision. One recipe for discovering *'whether a Woman will have the Man she wishes'* required the 'wearing' of two lemon peels all day, one carried in each pocket, then at night rubbing the four posts of the bedstead with them. 'If she is to succeed the person will appear in his sleep, and present her with a couple of lemons; if not, there is no hope.'[53] Even when such rituals involved a group of people rather than a single individual, secrecy and silence were often a feature. One involved a group of young women performing a silent ritual in order to see 'a future husband':

> On Midsummer eve, just at sunset, three, five, or seven young women are to go into a garden, in which there is no other person, and each gather a sprig of red sage, and then going into a room by themselves, set a stool in the middle of the room, and on it a clean bason [sic] full of rose water, ... then all are to set down in a row, in the opposite side of the stool, as far distant as the room will admit, *not speaking the whole time*, whatever they see, and in a few minutes after twelve each one's future husband will take her sprig out of the rose water, and sprinkle her shift with it.[54]

Although these books were sold in the public market place, they were clearly about very private rituals, and closely connected with female sexuality. Many entries have the sort of disguised reference to sexuality that the above references to lemons and to the sprinkling of shifts with water suggests. A popular recipe that can be found in one or another version in many different books of fate was for Dumb Cake:

In order to make the Dumb Cake to perfection, it is necessary to observe strictly the following instructions: – Let any number of young women take a handful of wheaten flour, (not a word is to be spoken by any one of them during the rest of the process,) and place it on a sheet of white paper; then sprinkle it over with as much salt as can be held between the finger and thumb: then one of the damsels must bestow as much of her own water as will make it into a dough; which being done, each of the company must roll it up, and spread it thin and broad, and each person must (at some distance from each other) make the first letters of her christian and surname, with a large new pin, towards the end of the cake; ... The cake must then be set before the fire, and each person must sit down in a chair, as far distant from the fire as the room will admit, not speaking a single word all the time. This must be done soon after eleven at night; and between that and twelve, each person must turn the cake once, and in a few minutes after the clock strikes twelve, the husband of her who is first to be married will appear, and lay his hand on that part of the cake which is marked with her name.[55]

This recipe can be understood to signify both active and passive interpretations of individual action. The sense of fate involved in all attempts to see the future may be seen as profoundly passive, but there is also a real sense of agency: in the action of baking, in the sharing of the ritual with friends who are willing to sit for hours in silence, and in the mingling of body fluid ('her own water') with the cake. Whereas cooking a cake might usually be thought of as the most traditional and conservative of home-making activities, here it has been transformed into an assertive source of power which promises to allow young women the amazing capacity to see the future. Similarly, in a recent discussion of knitting, based on Pauline Cummins' video installation *Inis t'Oirr/Aran Dance,* Catherine Nash notes that the making of an Aran jumper might easily be portrayed simply as a traditional West Ireland craft associated with reactionary views about the role of women. Yet in Pauline Cummins' work 'the distinction between ... active and passive which has devalued women's cultural production' is complicated with insights into the creative act of making the jumper, in which the sensuality of the knitting is explored. The artist examines the relationship of wool to body as a 'product of women's active, sexual narratives of desire ... [telling] stories of thought and touch', and describes 'the fantasy imposed on the man by the woman', an imaginary pleasure in tracing the lines of the wool, 'finger weaving'.[56] In the same way, the Dumb Cake mingles nurturing and demand, passivity and power. The fingers are the man's, as he places his hand on the part of the cake marked with the woman's identity, her initials and her water, while it lies still upon the virginal 'white' paper; but the preparation has been at the young women's initiative. It is they who 'salt' the mixture, adding piquancy, while perhaps calling to mind the many activities with which salt may be connected: crying, sweating, capturing, impregnating, sacrificing,[57] and it is they who prick the dough with a sharp new pin. Chapbooks like those which circulated this recipe, and which reformers criticised for a fatalistic erosion of enterprise, might be read instead as honouring a version of the domestic role of women in which their 'autonomous and active female sexuality' was acknowledged in ways that a more middle-class, rational culture either could not, or would not, recognise. The rituals were private and female, but the designation 'secret', as applied by the Society, expresses a disapproval and an assumption that 'private' also meant shameful.

Dreams and Rationality

Several middle-class writers were interested in dream books, and two in particular aspired to write them. H. G. Wells wrote:

> I figure to myself ... a sort of dream book of huge dimensions, in reality perhaps dispersed in many volumes by many hands, upon the Ideal Society. This book, this picture of the perfect state, would be the backbone of sociology. It would have great sections devoted to such questions as the extent of the Ideal Society, its relation to racial differences, the relations of the sexes in it.[58]

Mark Hillegas, who writes about Wells's fascination with the future, believes that in many ways Wells spent the rest of his life writing this dream book.[59] Like popular tradition, Wells's dream book was preoccupied with the future. In a lecture he delivered at the Royal Institution in January 1902, the author distinguished between two kinds of minds. The first, oriented to the past, always referring to precedents, regards the future as 'a sort of black non-existence upon which the advancing present will presently write events'. The second, oriented to the future, is constructive, creative, organizing. 'It sees the world as one great workshop, and the present is no more than material for the future, for the thing that is yet destined to be.' Wells's work was to predict what might be accomplished if the future-oriented mind were given freedom to express itself.

The other major writer who aimed to write such a book took a diametrically opposed viewpoint, regarding dreams as a revelation of the past, not of the future. It is in Sigmund Freud's famous dream book that we find rationality's greatest defeat of superstitious thought. The fact that Freud called his magnum opus his 'Traumdeutung', echoing the title of German chapbook dreamers, has been commented on by Harold Bloom.[60] In using the same title, Freud was linking his own 'interpretation of dreams' to the many widespread interpretations that circulated in popular 'dreamers', and it is reasonable to assume that he expected some comparison to be made. Not only was he aware of the genre, but he seemed to acknowledge a debt of inspiration to it: 'One day I discovered to my great astonishment that the view of dreams which came nearest to the truth was not the medical but the popular one, half involved though it still was in superstition.'[61] He rejected the scientific opinion advanced by writers such as Carl Binz, that dreams were froth ('Träume sind Schäume'), and was drawn, instead, to the popular view that dreams have a meaning, 'which can be discovered by some process of interpretation of a content which is often confused and puzzling'. Distancing himself from those who smiled at attempts to find a significance in dreams, Freud set out, with what we now know to be revolutionary consequences, to replace the ancient, traditional code of meanings with his own collection of symbols, a new lexicography of dreaming. By developing the role of dreams as fragments of suppressed wishes he appropriated the common dream book's emphasis on fear, hope, and desire; but by tracing those fragments back to childhood and repression, he transformed a significant aspect of nineteenth-century women's culture, moving the focus of dreams from the future to the past. The time-hallowed nature of dreams in chapbook literature, as oracles of the future consulted chiefly by women, was superseded.

Freud contrasted scientific discussion, which was interested in the source of dreams, and which generally suggested that they were an activity formed by physiological stimuli, and popular opinion, which was concerned not with sources but with meaning related to the prediction of the future, 'transforming the content of the dream as it is remembered, either by replacing it piecemeal in accordance with a fixed key, or by replacing the dream as a whole by another whole to which it stands in a symbolic relation'. The 'fixed key', which was the basis on which all dream books were based, became translated for Freud into a body of 'dream symbols' which, like dream books, could actually function quite independently of the dreamer's own contribution. '[W]ith the help of a knowledge of dream symbolism, it is possible to understand the meaning of separate elements of the content of a dream or separate pieces of a dream or in some cases even whole dreams, without having to ask the dreamer for his associations.' Freud believed that his approach was actually in tune with the 'popular ideal' of translating dreams according to a set of symbols 'used by the ancients'. He was simply updating the meanings of the symbols, which nevertheless he regarded as universal.[62]

Freud's work constitutes an appropriation of popular belief for the purposes of scientific discourse. I do not intend to claim him for an agenda of the sort carried out by social reform societies, but his re-interpretation of one of the most persistent strains of prediction in popular culture could only advance the cause of rationalism. By con-tributing to the replacement of dream books, with their representation of women's culture, his work facilitated a widespread acceptance of the dominant ethos of rationality as essentially 'masculine'. Its revolutionary nature was to take an influential trope in popular culture and to turn it around, so that by the early years of the twentieth century most educated people believed that dreams referred not to the future but to the past. The final paragraph of *The Interpretation of Dreams* refers again to popular belief: 'And the value of dreams for giving us knowledge of the future? There is of course no question of that. It would be truer to say instead that they give us knowledge of the past. For dreams are derived from the past in every sense. Nevertheless the ancient belief that dreams foretell the future is not wholly devoid of truth. But his future, which the dreamer pictures as the present, has been moulded by his ... wish into a perfect likeness of the past.'[63] In Freud's summary, the repre-sentative dreamer has been transformed from the woman for whom dream book compilers wrote to a universal man. Yet in clinical practice, Freud's emphasis on women perpetuated the common belief that women were a fertile source of repressed wishes. Freud's relationship to popular belief was one of both continuity and rupture: his insights into dreams built on centuries of belief that dreaming was connected with the innermost desires and fears, not solely with the discomforts of the body; but he broke with tradition in removing the interpretation of dreams from its important place in the private culture of women. In appropriating it for psychoanalytical investigation, he conferred important rational status, but removed an area of agency that generations of women had claimed as their own.

Conclusion

As with so much of popular culture and popular literature, the questions raised by books of fate far exceed our capacity to answer them. Why did they contain so little about friendship, or health? Why are the fortune-tellers consulted in illustrations and anecdotes in these books nearly always women, when most prosecutions seem to have been of men? One thing, however, is certain. Women's use of these books was not only a demonstration of gullibility. Those areas of life where some, perhaps mainly young women, experienced the most anxiety about the future are clearly signalled, as well as ways in which they tried to exercise some agency over implementing their dearest wishes. All who trusted to these books felt, although they may not have expressed it quite in this way, that the middle-class virtues of individualism and hard work were not enough to ensure against a potentially inequitable and harsh future.

4
Timeless Cultures

The Turk – who happily has much spare time – is constantly twiddling the lever of his cheap Austrian watch to keep it right. Nobody is ever sure of the time. The very fact that the Turks are satisfied with a method of recording time which cannot be sure unless all watches are changed every day, shows how they have missed one of the essentials of what we call civilisation.[1]

The historian Anthony Pagden has noted that whenever Europeans in the New World encountered anything that fell through their conceptual 'grid', they resorted to the terminology of the marvellous or wondrous: 'Such things existed out of real time and space.'[2] And, indeed, it often seemed to the European eye as if time itself was unknown in the new lands, with temporal difference featuring prominently in contemporary comment. Charles de La Condamine, for example, noted in 1745 that indigenous American languages were unable to express either 'time' or 'duration', and that the inability to express these (as well as other important concepts such as virtue, justice, liberty, discourse, and ingratitude) formed part of what he referred to as 'the little progress which the spirit of these people has made'.[3] Temporal conceptualisation was incorporated as part of the delineation of 'otherness' that was so essential to the creation of European identities.

British travellers abroad frequently resorted to a paternalistic discourse which displayed the same preoccupations as those which dominated the reform of popular culture within Britain itself. Temporal responsibility was taken to be an integral part of the maturity manifested by the educated classes, so that any difference in conceptualising time was described as a form of childishness. The infantilising function of colonial discourse has been widely documented and analysed:

The perception of the colonised culture as fundamentally childlike or childish feeds into the logic of the colonial 'civilising mission'

Figure 10: 'The Gypsy Fortune-Teller', *The Everlasting Circle of Fate* (1850)
This rather crude engraving displays a high-born young woman (with a
coat of arms on the door of her carriage) consulting a gypsy palmist. The
lady's companion, drawn in ethereal form, waits in the shadows. He may
be a figment of her imagination, or the ghost of a loved one. The gypsy is
accompanied by a black cat, suggesting witchcraft. Unlike some other
portrayals of magic, this image suggests that the artist had little understanding of the subject, with stereotypical toads and newts, and
hieroglyphics that seem to be mere squiggles, although the cauldron
hanging over the fire was a widely used symbol for sexuality.

which is fashioned, quite self-consciously, as a form of tutelage or
a disinterested project concerned with bringing the colonised to
maturity.[4]

The great British achievements in the area of time measurement
were, according to this view, a result of sophisticated scientific
knowledge, which required an education to be fully appreciated, so
that any difference in temporal practice was associated with a failure
of understanding. Above all, the field of astronomy epitomised the
benefits of scientific advances. James Ferguson, the famous
eighteenth-century populariser of astronomy, noted with approval

that Christopher Columbus had used his knowledge of predicting eclipses to overawe the superstitious natives of Jamaica.[5] In another example, the author of a popular early twentieth-century potboiler, *In the Power of the Pygmies*, made clear that it is his hero's superior temporal sophistication that allows him to exploit ignorance. Captain Crouch consults *Old Moore's Almanack* to find an eclipse, and then 'gathers the natives in the moonlight, chants a Victorian nonsense poem solemnly and, at the appropriate time, using a watch, tells the moon to disappear. The natives are terrified.' Readers are led to believe that eclipses had either never been seen in these distant locations or that no oral knowledge was passed on of their occurrence and meaning.[6]

The task of conveying knowledge to the colonies was frequently presented as synonymous with a pedagogical role for the 'parent' country. An important feature of this tutelage was the responsibility to convey a solid appreciation of time's steady progression and ultimately beneficial influence. In India, for example, 'the timeless stagnation of the Asiatic mode of production' was thought by some European commentators to be the fundamental cause of the country's poverty.[7] A host of European analyses of 'other' cultures, in both the nineteenth and twentieth centuries, placed temporal misunderstanding high on the list of the causes of backwardness and decline. For example, Anthony Aveni's history of the Mesoamerican calendar portrayed the Aztec's use of cycles in their calendar as an error with major political repercussions. This 'obsession perhaps unparalleled in the history of human intellectual achievement' contributes to a picture of Aztec society as fundamentally weakened, notably by a belief in the return of Quetzalcoatl.[8] Offering a less Eurocentric interpretation, Carol Greenhouse takes issue with Aveni's account, seeing the Western preoccupation with the Aztec leader's apparent misperceptions of time as a failure to understand the interrelationships of multiple descriptions of time, which in turn leads to a mistaken portrayal of 'a mentality incapable of grasping unique events'.[9] When scientific, European constructions of time are presented as the norm by which all other representations are measured, then anything which is not regular, linear, progressive and predictable presents an unacceptable aberration.

In contact with apparently unusual temporal observances, Europeans usually employed one of two categories. Some cultures were thought to be fundamentally lazy and irresponsible, avoiding the demands that precise time measurement would bring. Within

Britain, tramps and gypsies fell into this category – those groups which Henry Mayhew, in his journalistic description of mid-century London, dubbed 'the wandering tribes'. Others were considered to be totally unaware of the possibility of measuring time and its passing: to be, in a sense, timeless.

The Wandering Tribes

The first way of looking at unorthodox temporal practices, as a means of avoiding the discipline that might be required from accuracy and punctuality, has been the most commonly used category in the history of Western contact with other cultures. This supposed fundamental tendency to non-productive behaviour has also often been linked with the accusation of superstitious belief.

Allegations of 'superstition' to belittle minority groups has a long history. Protestant writers from the sixteenth to the eighteenth centuries represented Catholicism as profoundly superstitious, suggesting that this fault was closely linked to its fundamentally 'foreign' nature and its too close association with what might now be called 'cross-cultural' influences. The connection between the Catholic Church and the pagan practices which it had often sought to incorporate was roundly condemned, for example in an accusation that various Popes had been conjurers and sorcerers.[10] Indeed, Catholic ritual was so closely linked with magic and withcraft that in the reign of Elizabeth I the term 'conjurer' came to be synonymous with a recusant priest.[11] To critics, Catholic superstition was evident also in the use of time. The sixteenth-century Puritan divine William Perkins placed monks, mendicant friars and ultimately all Catholics in the category of rogues, beggars, and vagabonds because they added 52 saints' days to the 52 sabbaths, thus spending 'more than a quarter of a year in rest and idleness'.[12] Clearly a connection between false belief and the mistaken use of time preceded the nineteenth-century campaigns against fortune-telling.

Another large minority group which was felt by the middle classes to be irresponsible in the organisation of time consisted of those individuals and families who, for a wide variety of reasons, travelled in search of work. This included those who made a permanent life of tramping as well as those who travelled purposefully to find employment. Both categories were often looked upon with profound mistrust. While travelling from Lancashire to London on foot,

Samuel Bamford found one alehouse which refused to serve him with a glass of ale, despite his demonstrated ability to pay:

> A female of a superior appearance came and said they did not entertain foot travellers. I expressed my surprise at that, and assured her I was both able and willing to pay for whatever I called for. She said she did not doubt it, but it was an invariable rule of the house, not to serve persons travelling on foot, and the rule could not be departed from. Could I not have a draught of ale? I asked. No, foot travellers could not have anything there.[13]

This experience was not an isolated one, and Bamford takes the trouble to advise his reader, whom he clearly imagines is one of the working classes, how best to approach a lodging-house for a bed for the night:

> A foot traveller, if he is really desirous to obtain lodgings, should never stand asking about them. He should walk into a good room, – never into the common tap-room, – put his dusty feet under a table, – ring the bell pretty smartly, and order something to eat and drink, and not speak in the humblest of tones.[14]

'Respectable' dislike of a nomadic lifestyle had many causes, such as the fear of petty thieving, a simple dislike of the outsider, and the desire to protect the material resources of poor relief from undeserving claims; but this distrust was given rational form at least as early as the Putney debates of the seventeenth century when one side declared that unless a man had a fixed interest in the country (i.e. he owned property) he could not be counted on to exercise his vote in a responsible way. Two hundred years later Henry Mayhew suggested not merely rational but, more grandly, scientific grounds for mistrusting the itinerant. There was, he claimed, 'a greater development of the animal than of the intellectual or moral nature of man' in the nomad, and this might perhaps be due to 'a greater determination of blood to the surface of the body, and consequently a less quantity sent to the brain, the muscles being thus nourished at the expense of the mind'. Such a deficit of blood to the brain may, in Mayhew's hypothesis, have caused undesirable characteristics such as 'high cheek-bones and protruding jaws ... lax ideas of property' and a 'repugnance to continuous labour'.[15]

The 'wanderers' who were most closely associated with superstitious beliefs about time were the gypsies. Prejudice against them was not only associated with travelling, but also with their skin colour, their unfamiliar language, and their apparent irreligion.[16] However, perhaps the most famous feature of the gypsy stereotype was their skill in fortune-telling, particularly in the reading of palms. This could often be used to encapsulate the other prejudices as well, since fortune-telling could be considered a threat to Christian orthodoxy. One account of gypsy history claims that, after the first recorded visit of gypsies to the French capital, the Bishop of Paris excommunicated those who had had their palms read.[17] In 1803 prosecutions were brought against the well-known Norwood gypsies in London, a group whose reputation drew clients seeking fortune-telling services from miles around.[18] The charge against them was brought by the Society for the Suppression of Vice, and was for 'bringing idle persons about them at Norwood, to have their fortunes told on a Sunday'.[19] It is not clear which aspect of the offence was most repugnant to the Society, the fortune-telling, the sociability with idle persons, or the profanation of the Lord's Day.

The connection between superstition and gypsies was strong in the public imagination, so much so that there was considerable slippage between the travelling life, gypsy identity, and fortune-telling. Charles Godfrey Leland, antiquarian and scholar, declared that 'Gypsies ... have done more than any race or class on the face of the earth to disseminate among the multitude a belief in fortune-telling, magical or sympathetic cures, amulets and such small sorceries as now find a place in Folk-lore.'[20] The 'Egyptian' offence implied that the very act of claiming to be a gypsy was in itself a misrepresentation, presumably by ordinary rogues and vagabonds. In 1744 'all persons pretending to be gypsies, or wandering in the habit or form of Egyptians' were condemned, and linked with 'pretending to have skill in physiognomy, palmestry [sic], or like crafty science, or pretending to tell fortunes'. In 1824 a woman was arrested for fortune-telling but was actually prosecuted for 'pretending to be a gipsy'.[21] 'For most of the five and a half centuries that Gypsies have been in Europe', writes a recent historian, 'they have been lumped together with vagabonds and vagrants.'[22] In 1901 the *Termes de la Ley* referred to 'Egyptians, commonly called gypsies' who were 'counterfeit rogues, Welsh or English, that disguise themselves in speech and apparel'.

The deceit of passing oneself off as a gypsy fitted with the stereotypical representations of gypsies as deceitful overall.[23] By reading palms, for example, it was thought that gypsies were knowingly deceiving the gullible populace. A 1530 statute explicitly linked the dishonesty of fortune-telling with other forms of deceit:

> Diverse and many outlandish people calling themselves Egyptians, using no craft or feat of merchandise ... have gone from shire to shire and place to place in great company and used great and subtle means to deceive the people, bearing them in hand that they by palmistry could tell men's and women's fortunes and so many times by craft and subtlety have deceived the people of their money and have also committed many heinous felonies and robberies to the great hurt and deceit of the people they have come among.[24]

In reality, fortune-telling certainly was practised by gypsies. John Clare, for example, the visionary 'peasant poet' of the early nineteenth century, told of his encounters:

> In fortune telling they pertended [sic] to great skill both by cards and plants and by the lines in the hand and moles and interpretations of dreams but like a familiar Epistle among the common people that invariably begins with 'This comes with our kind love to you all hoping you are all well as it leaves us at present thank god for it' the preface to every bodys fortune was the same that they had false friends and envious neighbours but better luck would come and with the young that two was in love with them at the same time one being near and one at a distance —— one was a dark girl and one a fair girl and he lovd the fair girl the best etc

The formulaic representation of these predictions indicates Clare's mistrust and his belief that the gullible client was being cheated. He took some pride in his own ability to see through the gypsies' tricks:

> The credulous readily belevd them and they extorted money by another method of mutterd over their power of revenge which fright[ened] the honest huswife into charity —— I have h[e]ard them laugh over their evening fire at the dupes they had made in believing their knowledge in foretelling future events and trying

each others wits to see who could make a tale that might succeed best the next day[25]

Although many non-gypsies might believe that fortune-telling itself was a great trick, it was widely accepted that gypsies had amongst themselves many strange beliefs that would warrant the label 'superstition'. Those who admired gypsy life might see these as the accumulated wisdom of generations of oral knowledge; others might regard them as uneducated nonsense. Nevertheless, the portrayal of gypsy culture in the British periodical press underscored the representation of superstitious belief as closely connected to pessimism, warning, death, and foolishness.

In *The Virgin and the Gypsy*, D. H. Lawrence hints at the close connection between fortune-telling and a threat to rational timekeeping. A gypsy with 'dark, watchful eyes' invites a group of young people to have their fortunes told. The 'young ladies' are particularly tempted, but one of the young men, Leo, is uncomfortable:

> 'Don't the pretty young ladies want to hear their fortunes?' said the gypsy on the cart ...
> Oh yes! Let's!' cried Lucille at once.
> 'Oh yes!' chorused the girls.
> 'I say! What about the time?' cried Leo.
> 'Oh, bother the old time! Somebody's always dragging in time by the forelock,' cried Lucille.
> 'Well, if you don't mind *when* we get back, *I* don't!' said Leo heroically.[26]

The comment on Leo's heroism is no doubt ironic, but the consequences of throwing society's time observance aside are certainly portrayed as potentially dangerous. The close connection, too, between the young ladies' interest in fortune-telling and their nascent sexuality is as manifest here, in the early twentieth century, as it was in the Vice Society's representations of prediction at the beginning of the nineteenth century. Not only is Lucille physically drawn to the 'dark gypsy', but she has to withdraw into the murky interior of the caravan to have her fortune told, since it contains secrets too personal to be heard by the group outside. Sexuality, secrecy, young women's hopes and fears: Lawrence's portrayal of the role of the gypsies repeats all the themes expressed in the confrontation between Joseph Powell and the Society for the Suppression of Vice.

The compilers of books of fate also helped to maintain the tradition of gypsy superstition, particularly of gypsy women. As has been shown, many such books were attributed to a wise woman, even though their knowledge was depicted as passed on through an intermediary male author. Charles Leland's study of gypsy culture, assuming a grasp of an ancient oral culture that it could not possibly substantiate, notes that '[Gypsy] women have all pretended to possess occult power since prehistoric times.'[27] Mother Bridget, one of the common names to be found as an apparent source of nineteenth-century versions of cunning wisdom, was one of the 'Queens' of the Norwood gypsies. Bridget was a successful fortune-teller in the mid-eighteenth century, with a reputation for having accumulated considerable personal wealth from her clients.[28] Even when not mentioned by name, Bridget and her famous family of Norwood gypsies were shamelessly appropriated by chapbook compilers. *The Norwood Gipsy Fortune Teller* was one title, and *The Original Norwood Gipsy or Universal Dream Book and Magic Oracle* was another, the former published in Glasgow.

Also echoed in gypsy stereotypes were suspicions that prediction and gambling were linked. Late in the nineteenth century, while railway travellers were waiting for their train to arrive, they could pass an odd few minutes by putting a penny in what was called an automatic fortune-telling machine. *Pearson's Weekly* considered that its readers would have noticed such a machine not only at railway stations but also at many street corners.[29] In fact, it was really more of a gambling game, 'a handy roulette table', but there was a tenuous connection with speculation about the future. The face was covered with coloured triangles, which moved around alongside the figure of a gypsy girl, and the patron had to guess which triangle the finger of the girl would be pointing to when they came to rest. The term fortune-telling for this pastime had the loosest possible connection with the prediction of the future, but its appropriateness was reinforced by the figure of the gypsy. To try to know the future, it would appear, carried as little likelihood of success as a gamble.

Timeless Cultures

The second of the categories into which temporal difference was confined was that of apparently timeless cultures. Early anthropological literature is replete with declarations that some peoples were completely ignorant of time. In the 1930s E. E. Evans-Pritchard wrote of the Nuer, a tribe of the upper Nile:

> Though I have spoken of time and units of time, the Nuer have no expression equivalent to 'time' in our language, and they cannot, therefore, as we can, speak of time as though it were something actual, which passes, can be wasted, can be saved, and so forth.[30]

The Hopi of Arizona are perhaps the most famous example of a group of people who were, until quite recently, represented in Western academic literature as totally lacking a sense of time. Benjamin Whorf, writing at the beginning of the twentieth century, claimed that the Hopi had 'no words, grammatical forms, constructions, or expressions that refer directly to what we call "time", or to past, or future, or to enduring or lasting'.[31]

The role of the nascent nineteenth-century 'science' of anthropology in construing difference as lack has been much discussed in several recent studies, and a supposed temporal incommensurability in particular has been identified as a distancing technique used by some anthropologists. Johannes Fabian has suggested the existence of an analytical approach which he calls the 'denial of coevalness'.[32] This term highlights the way in which the producer of anthropological literature places the referent of anthropology in an 'other' time. The nature of some anthropological discourse as allochronic (Fabian's term signifying temporal difference) serves to exoticise other temporal practices, as the participant observer attempts to view them from an atemporal limbo.

Perhaps the most striking example of this 'timeless' category is Australia, where Europeans struggled to understand the new environment, but were oblivious to the temporal signs of the indigenous cultures. Aboriginal Australian culture was thought to be attuned to a temporal awareness quite other than that practised by colonial society. In the nineteenth century this supposed temporal 'otherness' was understood to be a total unawareness of time. The land itself seemed to have the potential to break free from European order. Numerous colonial records testify to settlers losing a sense of time: the seasons did not serve as a reminder of how the passage of the year was proceeding, imported clocks failed to keep going for long. The country itself was apparently timeless, and the timelessness of its people was deeply involved in their placement within a landscape that was disordered, threatening, and beyond recognition.

It was soon being noted that a general lack of precision in measurement was a handicap to Aboriginal people. A *Grammatical Introduction* to Aboriginal language in Western Australia noted that among Aboriginal people there was no way of expressing the adverb 'when', and that 'their mode of reckoning time is, by "sleeps" for short, and by the seasons for long intervals'.[33] In 1851 Bishop Rosendo Salvado of the New Norcia mission in Western Australia wrote that Aboriginal arithmetic 'stops at three ... anything more than that is *bulla* ... ("a lot")'.[34] This supposed inexactness was taken to result in a lack of concern about observing temporal schedules.

The opinion that Aboriginal people had little sense of the value of time has been repeated frequently throughout the history of European contact, right through to the present. In 1965 the Commonwealth Arbitration Commission included in its findings on a dispute about rates of pay the following comment on tribal Aboriginal culture:

> Time, in the Western sense, and the significance of time were ... unknown. [Aborigines] had no idea of forward planning, of working out a long term enterprise based on predictions of future planned occurrences. The notions of number, precise distance, and mathematical accuracy were unknown.[35]

The Commission used these findings, based on reports from anthropologists, to rule that Aboriginal workers were not entitled to equality of pay with white stockmen, under a slow-workers clause. In *White Man Got No Dreaming*, published in 1979, the eminent anthropologist W. E. H. Stanner writes: 'I have never been able to discover any Aboriginal word for *time* as an abstract concept.'[36] In present-day Australia it is a widely repeated nostrum of mainstream culture that there is a 'blackfella' time, one which is oblivious of (or resistant to?) white temporal schedules.

Several historians have suggested that the portrayal of Aboriginal culture as timeless was in the interests of European land claims. It was not new to represent lack of knowledge as disenfranchising a group of people. Anthony Pagden refers to the argument from colonial America that indigenous peoples, whose cyclical travelling colonists claimed was a form of nomadic practice, had undertaken no improvement and had thus forfeited their right to the land.[37] Nicholas Canny draws parallels between these claims and earlier English descriptions of the natives of Ireland, where categorisation of the Irish as nomads carried connotations of 'barbarism': 'In the travel literature that was read by sixteenth-century Englishmen nomadic people were considered to be at the opposite pole of civilization from themselves.'[38] In particular, the practice of transhumance, or moving from one pasture to another, was 'accepted as evidence that the Irish were barbarians, and the English thus satisfied themselves that they were dealing with a culturally inferior people who had to be subdued by extralegal methods'.[39] The doctrine of *terra nullius*, taken as justification for the British settlement of Australia, was founded on the belief that Aboriginal cultures had no space of their own, with a parallel construction of a lack of Aboriginal temporality. As always, time and space were intimately related, the absence of one a precondition for the absence of the other. Bain Attwood writes: '[A]boriginal space was declared to be a *terra nullius*, and was thus appropriated by the British, on the basis that its owners were conceived as belonging to a particular historical time or to no time at all (and thus prehistorical).'[40] The absence of territorial rights was a result of the invaders' belief that Aboriginal people were 'in the original state of nature', that of hunters and gatherers who had no concept of property.[41] Colonisation was seen as progress, the shepherding of an unchanging people into the benefits and duties of both 'spatial' development and temporal awareness.

In some accounts, Aboriginal timelessness has been presented as part of the nature of the noble savage, in tune with the cycles of the earth and the natural world, responsive to a more durable and 'real' temporality than the superficial measurement of the white man's clocks. In others, it has been about failure and inadequacy, an inability to adapt to Western society's goals of diligence and productivity. Both representations are useful if we are trying to understand the European cultures out of which they sprang, although it must be stressed that neither is an entry into an understanding of Aboriginal culture. Stereotypes of nomadic lifestyles, whether gypsy or Aboriginal, draw on European fears and beliefs.

For some, the supposed difference in the understanding of time has been a quality to be admired. Such a position is typified in Eleanor Dark's novels, written in the early twentieth century, *The Timeless Land* and *The Storm of Time*. 'Time was only the unvarying cycle of nature', she writes. 'There was nothing in his [the native's] life which spurred him on to change. Eternity was ever-present to him, past and future interwoven with his own life by legend and unvarying tradition, so that all time was the frame for his mortality, and contentment his heritage.'[42] Into this timeless and therefore unchanging Eden, the white man brought destructive knowledge, not of the difference between good and evil, but of the difference between the past, the present, and the future.

This construction of an Aboriginal culture steeped in mystical contentment still has considerable potency. Andrew Lattas has explored the use by mainstream Australian society of a 'romantic version of primitivism':

> the desire for the bush is often a wish to escape the endless march of time: 'Time is meaningless in this secret place ... Like animals carved on the rock, everything is very still.' ... The bush heals psychically by offering back to whites the spiritual truths taken from them by the superficial, material pleasures of cities. In particular, Central Australia is seen as a site of ancient, cosmic, healing power.

Lattas links this eulogy of the bush to the spiritual meanings which Aborigines find in the land: 'This theme of having one's emptiness healed by the fullness of Aboriginal culture is prominent in newspaper reviews of books and art exhibitions dealing with Aboriginal myths and paintings of the land.'[43]

However, such 'othering' of Aboriginal identity, discussed recently by Australian scholars as 'Aboriginalism', a practice closely linked to 'Orientalism', disempowers indigenous people by making them the object of European knowledge.[44] It elides, also, the mutual influence that such constructions carry, in which white expectations of mystical wholeness predetermine dialogue about Aboriginal belief, even on the part of Aboriginal participants. Furthermore, it carries the implication that Aboriginal power and identity lie in the 'undeveloped' world, leaving the centres of urban and commercial power to the white Australian.

Perhaps nowhere is the attempt to exoticise Aboriginal temporality so visible as in work on the Dreamtime. Recent analysis of the history of the word has stressed its hegemonic role within a dominant language. Did the word 'Dreamtime' accurately portray a phenomenon or world-view that was originally encountered by anthropologists in the Arunta word 'Alcheringa'? This would be to say that in accepting and using the 'English' word, Aborigines have recognised the empirical validity of an introduced term that filled a pre-existing 'lexical gap'.[45] Or would it be truer to say that in speaking of the 'Dreamtime' Aboriginal people are speaking English? Did the choice of the term spring out of the culture of the anthropologists who introduced it?[46] The present study contributes to this discussion not by looking at the various Dreaming terms and how they have been adapted since their introduction, but rather by looking backwards to the late nineteenth-century British connotations of dreaming.[47] The innocent transposition from one language to another of a cultural concept, especially one as complex as we are told the Alcheringa is/was, is unlikely. In the attempt to translate the incommensurable, the very word 'Dreamtime' itself played an important part in creating the connotations that now surround Aboriginal cosmologies.

There has been a close relationship between Australian society's privileging of 'traditional' Aboriginal culture and modern representations of the 'Dreamtime'. The question of what the 'Dreaming' means in Australia is complex, but there can be no doubt that for many it has come to stand for a mystical connection to the land and a challenge to modern, materialist practices. The opposition between the pre-modern world of the Dreaming and the world of white society was apparent from the term's earliest uses. In their 1904 glossary Baldwin Spencer and Frank Gillen defined 'Alcheringa' as the 'Name applied by the Arunta, Kaitish and Unmatjera tribes to the far past, or dream times, in which their mythic ancestors lived'.[48] This connection with the far distant past is one of the most widely repeated aspects of the 'Dreamtime'. When Bruce Chatwin referred to the Dreamtime in his popular and influential book, *Songlines*, he described it as an equivalent of the first two chapters of Genesis, a creation story set at the beginning of time.[49] However, by placing Aboriginal time concepts in the past, Australian society was able to suggest that indigenous peoples could not cope with change, and would therefore die out. The view that the Aboriginal race was doomed to extinction, unable to adapt to new conditions, was prevalent until the 1920s. In the same way as an Aboriginal history could not really be said to exist, since there was no record of change, so an Aboriginal future had no meaning. 'Aboriginal survival – an Aboriginal future, in other words – was a contradiction in terms ... [they] would soon no longer be (of the) present.'[50]

Far more common in the colonial record than a romantic eulogy of indigenous timelessness, however, was an attitude that unequivocally condemned it. For example, James Clow, a Victorian squatter, complained that a nearby tribe had left a sick old man without food or shelter, intending to return for him, but neglecting to do so for over a week. The old man, Clow writes, would have died had not the settlers intervened to save his life.[51] Here the lack of value attached to measuring time was said to cause such carelessness that it even endangered life. But some observers thought the phenomenon went deeper than this, suggesting an essential inability to understand the very concept of telling the time. Mrs Edward Millett, wife of an Anglican minister in Western Australia in the 1870s, noted in her journal that she had much difficulty in teaching a little Aboriginal girl, Binnahan, whom she had taken into her household, how to tell the time correctly. The girl quickly came to recognise the figures on the clock's face, and could relay an accurate description of the relative positions of the two hands, but she still could not tell the time.[52] The implication of cultural or racial handicap is pronounced, especially since Mrs Millett had earlier attributed a widespread inability to tell the time among the colonists themselves to the scarcity of clocks, not to personal ineptitude.[53] Binnahan's eye and mind were quick, and there was no doubt about her intelligence, but, according to Mrs Millett's account, she seemed unable to grasp the concept of measuring time. Although Mrs Millett did not disapprove in any obvious way, and indeed would surely have considered her approach to be one of Christian charity and generosity, her own cultural inability to accept time consciousness as a human universal contributed to Binnahan's marginalisation as surely as James Clow's more obviously disapproving attitude.

Another well-known term that is closely associated with Australian Aboriginal culture is also related to temporal practices. At least 70 years before 'Dreamtime' was first used, British observers were applying another time-conscious word, 'Walkabout', to the aspect of Aboriginal life that most frustrated and annoyed them. It, too, suggests a lack of awareness of temporal regularity, although it was not expressed overtly as a judgement about time. Both the Dreamtime and Walkabout are intimately linked in a terrain of temporal 'otherness' that was first mapped in nineteenth-century rationality. In fact, in the *Oxford English Dictionary* the word 'walk' has a sub-entry which says that to walk about is to 'perform ... actions as a somnambulist'. There is an aimlessness implicit in 'Walkabout' in popular British usage. If a possession goes walkabout, it means it has been lost or mislaid. It is no longer fulfilling the purpose for which it was intended, and is, in effect, useless; just like a sleepwalker it is moving in suspended animation, unrelated to real time.

The term appeared in print for the first time in the *Sydney Gazette* of 1828, reported as being used by an Aboriginal man, Tommy, who was executed for murder. As he stood on the scaffold, he said, 'Bail more walk about', and the newspaper translates him as meaning that 'his wanderings were all over now'.[54] He was given to using unusual words, the paper tells us, that had not previously been heard in the Sydney district. The implication is that he had come into the area from far away, bringing terminology that was new. However, responsibility for the negative connotations of this word cannot be attributed to indigenous people. There is ample record that other words, which could have become the dominant usage, were being used elsewhere. Alfred Joyce, in reminiscences of a pioneer life, commented that in attempting to communicate their travelling to the white man, Aborigines chose the words 'Pull away'.[55] This seems just as memorable and convenient a term for nomadic practices as Walkabout, and would, perhaps, have conveyed an Aboriginal viewpoint more accurately, since it conveys a sense of direction.

The term 'Walkabout' is a reminder that at the very heart of representations of indigenous people as timeless is a deep abhorrence within Western European culture of the practice of nomadism. Aborigines were accused of not being willing to work regularly, of being unpunctual. They simply would not observe the temporal regulation that their employers required of them, because they kept moving away from their settlements. Alfred Joyce wrote that Aborigines were given plenty of space, but that there was 'no possibility of getting them to stay on the place for any length of time'. Despite having many material advantages to encourage them to stay (cattle and sheep, farming implements, and clothing) 'most would wander off again after a week or two and none would engage in any industrial pursuit'. He deduced that 'their nomadic habits were the great difficulty in any systematic plan of settlement on their behalf'.[56] Joyce's chapter on Aborigines begins with a description of what he calls a 'peregrination'. The use of a term that is based on a sense of being a stranger in a strange land, a voyager, distances indigenous people from their own environment, suggesting that they were alien, foreign to it.

Critics felt that wandering was an indulgence, an escape from useful employment, and an excuse for wild behaviour. In 1853, for example, Hugh Jamieson wrote to Bishop Perry, Anglican Bishop of

Melbourne, that the local Aborigines had no wish to abandon their unsettled and roving life and that during periods of annual migration, 'The tribes ... slowly and indolently saunter along the banks of the Murray and Darling.'[57] However, Jamieson's use of the words 'slowly and indolently' are a key to another important issue underlying this antipathy. It was the apparent lack of productivity of Aboriginal 'sauntering' that seemed to offend. Indeed, conflicting attitudes to time underpinned and continue to underpin Western distaste for nomadism. It was for this reason that what was seen as the productive travelling of the itinerant bush worker was considered totally unlike Aboriginal nomadism. Cultural preconceptions and political advantage combined to make British colonists and administrators blind to a form of knowledge in which material 'improvement' and the accumulation of property were not primary concerns.

The Future

But most important of all, at a time when belief in progress and the benevolence of the future were central to national pride, British travellers disapproved of 'primitive' fear of the future. As shown earlier, the terror brought on by eclipses became a cliché in imperial fiction, as a 'convenient and dramatic symbol of the gullibility of primitive peoples'.[58] Although accounts vary as to why 'native' peoples were afraid of eclipses, they are often represented as fearful that the darkness will last. Where contact with technology or dread of celestial events demonstrated fear, the observer assumed a fear of the future itself.

As with other aspects of cross-cultural contact, representations of 'native' ignorance of technology, fear of celestial phenomena, and dread of the future had its roots in European society itself, in which prediction was most usually identified with warning. Street literature exhibited persistent motifs of forthcoming doom. Prediction and death were interlocking themes. The post-1736 Parliamentary focus on fraud did not replace a popular emphasis on death as one of the most fascinating parts of the whole business of consulting a fortune-teller. In 1815, for example, the *European Magazine* recorded that Mrs Spaul, 'a pretended fortune-teller', had been committed to gaol as a vagrant, not because she had received payment under false pretences, but because her predictions had frightened her client to death.[59] The distinction between educated confidence and timorous gullibility, so prominent in colonial contact literature, was first drawn in descriptions of popular belief within England itself. The uneducated fear of the future was portrayed as the working classes' inability to see that their own actions in fact created the future, that planning was the essence of optimism. Mayhew's description of the nomad stated that he could be 'distinguished from the civilised man … by his inability to perceive consequences ever so slightly removed from immediate apprehension'.[60]

Accuracy of prediction was the most important function of temporal sophistication. Even where the anthropological eye could see no truly 'scientific' measurement of time, the ability to forecast accurately was admired. For example, despite the lack in temporal understanding which was an accepted feature of Hopi society, the historian G. J. Whitrow noted that 'the Hopi have successfully developed an agricultural and ceremonial calendar, that is sufficiently precise for particular festivals seldom to fall more than two days from the norm'.[61] In other words, an impressive (and perhaps surprising) degree of predictive accuracy had been obtained, despite a lack of 'real' knowledge. Whitrow's account, published as recently as 1988, demonstrates the Western presupposition that the primary purpose of the calendar is to make the year predictable. It should be noted, in passing, that the work of Ekkehart Malotki has demonstrated that the representation of the Hopi as having no temporal discourse was mistaken,[62] and that they have (and had in the past) complex conceptualisations of time.

If one of the most important functions of Western temporal technology is to make the future more predictable, in the sense of more regular and amenable to planning, then the accusation that 'timeless' peoples are less able to cope with change than Western, industrialised peoples is strange, since an attempt to predict the future is an attempt to decrease the likelihood of unexpected change. There is an inherent inflexibility in punctuality. Cultures which come increasingly to rely on technological regularity, which they may call time, are actually attempting to limit the possibility of unplanned developments. The future becomes a challenge, one that must be met in combative mode. It is this encounter with the future that lay at the heart of many expressions of temporal misunderstanding during the establishment of European colonial supremacy. Colonists did not countenance the possibility that some indigenous cultures' apparent lack of planning might have indicated an inherent trust in the future.

In *The Timeless Land*, set in early colonial Australia (though written in 1941), differing attitudes towards the future are fundamental to the gulf between European and Aboriginal. Governor Phillip's excitement as he contemplates the future of the new land is contrasted with Barnagaroo's premonition of frightening things to come, including death for her people. Phillip has an intuitive insight about the future greatness which will follow the humble beginning of the colony. Indeed, it is one of his main qualities as a leader that he possesses the capacity to envisage a great future. Yet this self-validating belief manifested by Governor Phillip does not suddenly surface on contact with indigenous people in Australia. It is a quality that also sets him apart from his own countrymen. The convicts, in keeping with their less-educated horizons, are unable to share any enthusiasm for the future. He does not expect them to share his dream, only to co-operate in the present.[63] The gulf between the rulers and the working classes is as wide, as far as faith in the future is concerned, as that between coloniser and colonised.

5
Calendar Girls

One of the more surprising conclusions of a study of time practice in the eighteenth and nineteenth centuries is that women seem to have possessed a certain temporal authority. They had traditionally been the purchasers of calendars, as well as the imagined readers of their literary content. The male publishers and compilers were perfectly aware of this timekeeping element of women's role. As I have discussed elsewhere, the most popular almanac, or calendar, of the eighteenth and nineteenth centuries was *Moore's*, which was frequently portrayed as being for women, and of which *Notes and Queries* wrote: 'if anyone ventured to dispute its accuracy in the presence of an old woman with a mop in her hand, woe be to him.'[1] The most successful new calendar of the eighteenth century, the *Ladies' Diary*, subtitled the *Woman's Almanack*, was specifically intended for the 'Use and Diversion of the FAIR-SEX', with articles on 'female virtues', the nature of love, recipes, and 'A Brief Chronology of Famous Women'.[2] In 'reforming' attitudes to the future and to forward planning it was desirable either to remove temporal authority from women (by denying their beliefs in prophetic dreams, for example), or to inculcate in the woman reader a different expectation of her role. It is the second of these processes that this chapter examines, by exploring how the gendered nature of the calendar became closely linked with the gendered image of the nation. The portrayal of women on calendars became a means of promoting the idea of progress, strengthening the message that the care of personal time would in turn contribute to the nation's well-being.

The late nineteenth century saw increasing use of public space to promote national prestige. This was most frequently done in the form of statues and monuments, such as those of Queen Victoria, orb and sceptre in hand, which looked down from park and city centre throughout the British empire. Across Western Europe, similar symbols of national identity proliferated. Most had certain features in common: first, a wish to claim a sense of the past, of a tradition

Figure 11: 'Arid August', *Punch* (1884), Battye Library
One of the earliest depictions of a glamorous young woman as a symbol for the passage of the year. Her sultry demeanour links her with the rise of the late nineteenth-century figure of the femme fatale.

which conferred authority and legitimacy; second, traces of revolutionary symbolism, using iconographic female figures of liberty or winged victory; and third, references to cornucopia, indicating the prosperity that nationhood would bring. Such statuary was fundamentally triumphalist, celebrating the achievement of unity and reassuring the citizenry that this promised a bright destiny.

At the same time, images of women also began to appear in British calendars. Since the calendar's usefulness depended on its being widely accepted by the community, it might in some senses be considered a public space. Some calendar illustrations certainly played a similar role to public monuments, with Britannia and other figures of national pride (the lion and the jolly sailor), but there were also messages linking the female form specifically to temporality. This chapter argues that, in taking the symbolism of nation into the home, the new calendars of the late nineteenth century contributed to the overlapping of public and private time, as well as reinforcing links between nationalism and temporal responsibility.

It has been claimed that works of public commemorative art help to manipulate memory in order to create consent. Certainly they have usually been intended to contribute to a sense of identity based on a shared symbolic discourse.[3] The calendar, too, was seen as creating community, with its festivals and weekly cycles. In fact, all three of the characteristic features of public statuary could be found in calendars. A sense of the past was exactly what, in their role as remembrancers, they had long perpetuated, with chronological lists of important anniversaries; icons of liberty were perfectly situated when related to the dates of revolutionary events which they exemplified; and, finally, the cornucopia theme could hardly be better expressed than in a seasonal image, with the fruits of harvest bundled together, preferably in the arms of a woman signifying fertility. These shared motifs linked the symbolism of nation to the symbolism of the passage of the year.

A similar iconographic purpose had for some time been served by the image of 'woman as nation' on currency. Britannia had appeared on every banknote issued by the Bank of England since 1694, and was probably the most widely recognised female figure in the country. Victoria herself may have rivalled this level of recognition, since she appeared not only on currency but also on the newly established penny post. In fact, the figures of Britannia and Victoria were sometimes merged, to produce a Britannia who looked strangely like the monarch. Both figures combined maternal dependability and wisdom with authority and strength.

However, symbols of nation and symbols of the year were not completely interchangeable. The issue on which they most clearly diverged was in representations of progress. The figure of Victoria could convey messages of power, stability, and prosperity, but as she grew older, her appearance was less likely to convey hope for the future. Even while she was relatively young, she withdrew into seclusion and mourning for Prince Albert, and, after she was drawn back into public life, retained the black clothes of a widow. While increasing age and sombre appearance were no handicap for the figures of monarch or Britannia, as Victoria entered her sixties pictures of younger women also appeared on the scene. These younger images conveyed a message of progress and abundance with no hint of age and decline.

Young women were particularly likely to appear on calendars, where they stood for the new year. Some calendars experimented with a young child as the new year, but the future as young woman was far more successful. This was, no doubt, related to themes of fertility and sexuality. A young woman could be both virginal, like the new year, but also fecund, as her placement amidst seasonal imagery attested. She was often draped in classical dress, echoing the widespread use of Greek and Roman legends in late nineteenth-century British art and literature. According to one analysis, 'the images of women produced by archaeology and mythology shaped the consciousness of the nineteenth century'.[4] It has been suggested that classical myths were used in ways that emphasised women's sexual identity: 'These ideograms existed to be decoded by a patriarchal society well informed about classical mythology ... [W]oman's biological function was construed as central.'[5] The artistic use of nymphs, for example, whose mythical sexual activities were reflected in the development of the word 'nymphomania', was often a component of these tableaux. Fertility as beauty has long been an enduring theme in art.[6]

However, although biological and sexual messages were certainly an important part of the appearance of these nineteenth-century calendar girls, there were also clear messages about youth's optimism. The calendar was one of the few places where the young and courageous New Woman of the eighties and nineties was represented in pictorial form. There were plenty of literary references to this new manifestation of emancipated womanhood, but in paintings she was 'very elusive indeed'.[7] Lampooned as much as admired, nevertheless the qualities of energy, intelligence, courage, and strength were features that the New Woman (in her various guises) brought to the forefront of public discussion, and features that in turn found their expression in representations of progress.

To some extent, the increasing appearance of images of young women in late nineteenth-century periodicals can be traced to technological development. Throughout the century, new printing techniques brought a decrease in the cost of illustrations, combined with improvements that allowed complexity of line. Woodblocks were replaced by wood engraving, then steel engraving, lithographs, and, at the very end of the century, half-tones based on photographic plates. This cheaper illustrative material was seized on by the flood of cheap calendars which had appeared on the market following the abolition of restrictive publishing taxes. The sale of calendars with which to organise busy Victorian lives became a lucrative business, on which rode the fortunes of Cassell and Letts. But in an overcrowded market, with calendars selling for as little as a penny, how were purchasers to be tempted to buy? The answer lay in illustrations. Authors and publishers alike considered that illustrations were crucial in influencing people's choice of purchase, with some claiming that people would not buy anything printed at all unless it had pictures.

The calendar itself was high on the buying public's list of priority items. In 1841 its astonishing popularity saved the life of the new comic periodical, *Punch*. Mark Lemon, Gilbert à Beckett and a small group of colleagues had attempted to find a place for a journal in a sector of society that had not previously been targeted – those members of the new middle classes who had had enough education to understand an occasional classical reference but who were critical of the hierarchies of inherited privilege and power. In providing satire, both social and political, *Punch* hoped to cultivate a new reading public. Although we may now think of it as an outstanding literary success in periodical publishing, this bold attempt almost failed. By the end of 1841 the journal was on the point of closing. Henry Mayhew, the journalist who perhaps more than any other prided himself on being close to the tastes and interests of ordinary people, suggested that *Punch* might provide a free almanac, or calendar, with the first issue of the year. In January 1842 *Punch's Almanack* was launched, and circulation went up in one week from 6,000 to 90,000, an increase 'unprecedented in the annals of publishing'.[8]

Punch's calendar is an invaluable source for changes in ways of thinking about the passage of the year. Aiming for a relatively popular market, it was amply illustrated, with many of its drawings expressing the same tone of gentle satire carried within the rest of the publication. Its images of women track themes about the passage of time, and the relationship of time both to gender and to concepts of nation. Susan P. Casteras, writing about representations of women in Victorian art, comments that

> Throughout the entire period it was more often in the realms of periodicals like *Punch* that women received more 'realistic' treatment, for in spite of its satirical undertones, *Punch* at least humanized them as fallible beings and not paradigms of perfection.[9]

Punch is indeed an extremely important source of illustrations representing the social history of women's lives, not least because it is the repository of so many thousands of black and white drawings; and for the majority of *Punch* illustrations, it would be true to say that the artists took pains with realistic detail, of dress, lifestyle, and surroundings. However, within the calendar the young women were entirely paradigmatic. Their iconic status is revealed not only in the plethora of classical imagery and political allusion, but also in their sometimes quite blatant sexuality. The 1880s, the very time when prominent calendar illustrations of young women began to appear in *Punch*, were also the decade when the social purity movement was in full swing and in which particularly fierce public debates took place about the morality or otherwise of nudity in art. Outraged comment followed several of the Royal Academy exhibitions of that decade.[10] *Punch*'s artists seized on the opportunity of depicting woman as the passage of the seasons, and therefore as nature, using poses and draped clothing that made the female body as 'natural' – as sexual – as they could. Their calendar images came close to suggesting nudity, with each month's representative either locking gaze provocatively with a male observer, or else withdrawing with a coy raising of one knee in mock defence. *Punch*'s calendars displayed a tension between the acceptable use of youth to denote progress, and the potential for transgression, in the artists' use of the young, female body to flout current conservative strictures.

The use of ever-younger women to stand for the year also underscored messages about ephemerality, and this was not always reassuring. Whereas the coming of the new century had gone largely without comment by calendars in 1801, towards 1901 an obsession with the rapid pace of change increased not only self-congratulation about technological progress but also anxiety. In 1887, the Royal Academician Sir William Quiller Orchardson exhibited a painting called *The First Cloud*, in which a wife appears to be on the point of leaving her husband, a not uncommon theme in late Victorian art. *Punch* parodied this dramatic moment by producing its own drawing of the same scene, with the husband, in artist's smock, addressing a departing young woman as if she had just sat as his model, perhaps for a calendar, combining messages about the changing year and the pace of moral and social change:

> Yes, you can go; I've done with you, my dear.
> Here comes the model for the following year.
> Luck in odd numbers – Anno Jubilee –
> This is divorce Court Series Number Three.[11]

One of the earliest images of young calendar women was in the *Punch* of 1884, when the artist Linley Sambourne portrayed August as a young woman sweating in the heat of the midday sun (Figure 11). In modern beachwear, she is nevertheless clearly a symbol rather than a realistic representation. She is surrounded by images that make no claims on 'realism': Gladstone as a bee sucking at the flower of a Scottish thistle, a frog bearing a tricolor and a rifle, grouse shooters in the hills, and Father Thames reduced to a skeleton since his water supply has dried up. She herself, with long black hair trailing down her back to her knees, could well be a water nymph, for all the modern setting. The spray reaches the spot where she is sitting on an unseen rock, and rises up through the curve of her arm. Most sensuous and nymph-like of all, though, more than the raised right knee, the sultry, downcast eyes and the bare toes, are the almost bare breasts, revealed by a neckline that is not open by dressmaker's design but rather thrown aside to cope with the unbearable heat. This is a nymph who has been transformed into a femme fatale. The powerful emergence of the femme fatale, this new 'shadow' side of Victorian femininity, has been traced by Susan P. Casteras to the 1870s. The quintessential femmes fatales described by Casteras were 'voluptuous, often with abundant tresses ... both indolent and withdrawn in private reverie ... communicating a sense of magnetism and even potential danger with their mystical silence, half-closed eyes, sensuous mouths, and hypnotic looks of anes-thetized expectancy'.[12] This is a more than adequate description of Sambourne's 'August'. The admission of this near-pornographic image into middle-class homes was effected under the guise of rep-resenting the calendar.

Four years later, under the rubric of 'A Classic Calendar', a title that would allow ample reference to the sexual motifs of classical mythology, Sambourne (Figure 12) repeated several of the themes beloved by late nineteenth-century art. Flora, probably modelled on the same woman as in August 1884, holds out a spring flower to indicate the month of May, while above her Danae shelters from Zeus's golden shower of rain, and below Hera, Athena and Aphrodite compete to win the golden apple from Paris. *Punch*'s laconic tongue in cheek is conveyed by the anachronistic umbrella sheltering Danae and the strangely hybrid Paris with his Greek legs and urban top hat and monocle. It may be too fanciful to draw a link between Sambourne's use of Danae and her central importance to the notorious series of articles in the *Pall Mall Gazette* by W. T. Stead, *The Maiden Tribute of Modern Babylon*. These articles, claiming to expose the 'white slave trade' that provided young girls to the brothels of England and Europe, had appeared in July 1885 and attracted enormous public attention. The myth of Danae imprisoned by Zeus and impregnated against her will by his trick of appearing as a shower of gold, became a metaphor for prostitution,[13] one that would have been widely read and discussed just a few years before the appearance of this calendar. At the very least, *Punch* was capi-talising on the currency of the myth in public discourse.

In 1899, *Punch*'s calendar was still showing young women, but the style had changed. Now (Figure 13), under the pen of Bernard Partridge, the young women are modern, with bold gazes, some standing hands akimbo as they queue to be presented to Mr Punch. They clearly belong to the New Woman category, rather than that of the femme fatale, with many exhibiting an independent and wealthy lifestyle as they prepare for some smart event, such as grouse shooting, the regatta at Henley or the Epsom races. Indeed, the scenario, in which each is being presented to a Mr Punch who seems to be the host of the occasion, calls to mind one of those débutante balls that became fashionable amongst the rich in late nineteenth-century Britain. In 1900 the journal *Queen* began to carry portraits of the year's débutantes, indicating increasing public interest in the 'season'.

Figure 12: 'A Classic Calendar, or Myths for the Months',
Punch (1889), Battye Library

April, May, and June. The traditional calendar associations with harvest
and fertility are linked with young women who, in their loose-fitting
classical robes, bare feet and long flowing hair, seem very different from
the corseted and constrained models of other Victorian periodicals.

Figure 13: 'Mr Punch's Almanack 1898–1899', *Punch* (1899), Battye Library
Earlier calendar images of the zodiac have been transformed here into
twelve young women, each carrying astrological signs and appropriate
props to suit her month. A thirteenth maiden symbolises the year as a
whole, with a question mark for head-dress. The Twelfth-cake that Miss
January carries is a yuletide practice that seems to have faded. 'The costly
and elegant Twelfth-cake,' writes Robert Chambers in *The Book of Days* (vol.
1, p. 64) was a highlight of Twelfth Day (6 January). 'Formerly, in London,
the confectioners' shops on this day were entirely filled with Twelfth-cakes,
ranging in price from several guineas to a few shillings.'

In Partridge's image of the new year, the usual themes of fertility are present, but they have moved to an interior context. Rather than fields of corn, we have here a drawing room, with curtains closed to exclude the natural world, and only rose petals festooned at the feet of the young maidens and garlands of fruit hanging from the walls to hint at the theme of abundance. This perhaps reflects the 'enormous burgeoning of interest in the appearance and decoration of the home' in the second half of the century.[14] Household decoration was not only an expression of middle-class prosperity, but also a beneficiary of technological advancement in products such as wallpaper and tiles, an area especially beloved of the Arts and Crafts movement. There was also a move within Victorian narrative painting towards more 'bourgeois subject matter ... more earthbound and domestic',[15] and *Punch* was, of course, very much part of the Victorian art scene, many of its illustrators having been distinguished artists in their own right.[16] However, the move indoors may also simply be a more explicit extension of the sequestered environment of earlier outdoor scenes. The garden was a segregated space, and as such could indicate 'female innocence and unavailability' as much as the drawing room or parlour. The 1889 calendar (Figure 12), in which Flora offers a flower to an unseen admirer (the two classically robed men are standing too far away to be the recipients of her generosity), has a clearly delineated frame within which Flora sits, rather like many contemporary drawings of young women in a garden setting. The later, more urban ingenues of the drawing room were in a line of continuity with the pastoral maidens of the mid-century.

The thirteen young women lined up in the drawing room of 1899 may well be the first example of the use of a calendar girl to represent each month. While the young maiden at the head of the line represents the year in total, and displays a question mark on her head-dress to symbolise speculation about the future, the artist retains individual zodiac references for each month. February shows not only the hearts of St Valentine's day on her postal delivery bag, but also the fishes of Pisces on her breast, and the fool of April carries a wand with a Taurean bull's head. The artist's palette carried by 'May' displays the initials of the Royal Academy, a reference to the annual exhibition, yet another indication of the close connection between *Punch*'s illustrators and the 'respectable' art world.

The theme of presentation to Mr Punch is a continuing one. Miss 1900 (Figure 14) is clearly labelled as a débutante, and is in the

Figure 14: 'The Débutante', *Punch* (1900), Battye Library
Here the image of the year and of the nation begin to merge.
The demeanour of the young new year is confident as she steps
out with the world (particularly the East Indies) at her feet.

THE
NINETEENTH
CENTURY
AND AFTER

No. CCLXXXVII—January 1901

This Janiform head, adapted from a Greek coin of Tenedos at the request of the Editor, by Sir Edward J. Poynter, P.R.A., tells, in a figure, all that need be said of the alteration made to-day in the title of the Review.

Figure 15: *The Nineteenth Century and After* (1901)
The first twentieth-century edition of this prestigious periodical,
begun in 1877, which had been called simply *The Nineteenth Century*.

process of being presented to her host. Her appearance follows the instructions to débutantes from the pages of *The Lady*: 'Everyone must of course wear plumes and lappets in the hair.' Her dress seems vaguely classical, and is, no doubt, white: 'Young girls ... to be presented at a Court Levée must wear white ... the train being kept to one side by bunches of ribbons and white flowers.'[17] The shell around her neck recalls Botticelli, who, at the end of the century, became 'a valuable part of an ambitious late Victorian's cultural capital', though this rapidly became a 'vulgarization', with 'the "Primavera" ... in most drawing-rooms of the suburbs'.[18] The setting is again an interior, with the garlands around the curtains showing fruit and fertility. The young woman's gaze is bold and confident, but firmly linked to her class and status. She may be flirting with Mr Punch, whose look is humorously lascivious, but she is looking down on him. She is an embodiment of British economic success: at her feet lies the world, displayed as a map woven into the carpet. This is a young Britannia, full of confidence and power, depicted in a much more sexualised and class-conscious way than ever before.

The new year as woman was not confined to calendars. In January 1901 the leading periodical, *The Nineteenth Century*, also chose to represent the new century with the image of a young woman (Figure 15). In this Janus-like figure, the nineteenth century, passing away, is represented as a distinguished, Socratic old man. The journal, after all, had been one of the most successful serious periodicals during the last quarter of the old century. However, the old man has a downcast look, showing that his day is over, and that he is tired. The future belongs to the bold young woman, with her upright gaze and confident thrust of the jaw.

The question of why the future in this image should be presented as a woman rather than as a young man or a child brings us back to the imagery of the nation. In their work on 'The French Revolution and the Avant-Garde', Ian Small and Josephine Guy write that the influence of the French Revolution – the revolutionary metaphor – was a necessary precondition for the avant-garde, the group of British artists who in the 1870s 'transferred the revolutionary critique into the domain of artistic forms'.[19] We begin to see calendar images of women at a time when Small and Guy say nearly all British avant-garde writers and artists ardently propagandised the revolutionary heritage of their French counterparts. The use of the young, female, classical body speaks of fertility, but linked with French iconography it also speaks of liberty and the nation, tropes that came together in

Figure 16: *Liberty Guiding the People*, Eugène Delacroix (1831)

France in the figure of Marianne. Woman as liberty was indeed fierce and strong, but she was also the mother of the nation, nurturing and fertile. The most famous and influential depiction of this symbol is Eugène Delacroix's *Liberty Guiding the People* (Figure 16). Here the woman as liberty strides with Amazonian fierceness across the bodies of fallen patriots, with a rifle in one hand and a flag in the other. Her head is turned to profile, suggesting both Delacroix's strong interest in antique coins and her own emblematic status.[20] The young woman in *The Nineteenth Century*'s cover also has her head turned to profile, and is indeed adapted from a coin, as the caption below it indicates. Coins were, in fact, the most frequent site for images of Janus, and Sir Edward Poynter, the President of the Royal Academy, had used the coin image elsewhere, a simple 'classicising' technique.

The image of woman as both classical goddess of victory and also as madonna-like mother of the nation has been vividly described by Marina Warner, who explains the importance of the slipped chiton, the classical robe falling carefully loose to reveal just one bare breast:

> By exposing vulnerable flesh as if it were not so, and especially by uncovering the breast, softest and most womanly part of woman, as if it were invulnerable, the semi-clad female figure expresses strength and freedom. The breast that it reveals to our eyes carries multiple meanings, clustered around two major themes. It presents itself as a zone of power, through a primary connotation of vitality as the original sustenance of infant life, and secondly, though by no means secondarily, through the erotic invitation it extends, only to deny.[21]

Marilyn Yalom too writes of the breast as an image that was carried into nineteenth-century culture bearing traces of eighteenth-century symbolism:

> At no time in history – barring our own age – have breasts been more contested than in the eighteenth century. As Enlightenment thinkers set out to change the world, breasts became a battleground for controversial theories about the human race and political systems. Before the century was over, breasts would be linked, as never before, to the very idea of nationhood. It is not too far-fetched to argue that modern Western democracies invented the politicized breast.[22]

Although the bare breast could not be freely used in British national iconography during the Victorian period, nevertheless artists found ways of referring to this same aspect of symbolism. In Figure 14, while the young débutante is virginal and demure, the *art nouveau* candelabra that takes almost central place in the composition is allowed an Amazonian bareness of breast, uniting with the 'real' woman's youth and strength of purpose to emphasise

Figure 17: 'A Garland for May Day', Walter Crane (1895)

messages of fecundity and courage. Attention is drawn to the débutante's bosom, which is banded about by the number of the year. Her sexuality and fertility are further emphasised by the vase-like clasp lying across her lower body, and by the flirtatious look she gives Mr Punch. Only the wing-like sleeves on her dress remind us that she is not meant to be taken as real, but is rather a symbol of the nation, even a symbol of destiny itself.[23]

Perhaps the most famous example of the overlap of France's revolutionary motifs with Britain's national symbolism is the picture of the heroine carrying sheaves of wheat in Walter Crane's design for May Day 1895 (Figure 17). Crane's 'woman as nation' bears a close resemblance to Marianne, particularly in the use of the Phrygian cap, but the banner flowing around her feet carries a message that 'England should feed her own people'. Crane's socialism explains his appropriation of French iconography, but the image also carries links with May Day's pagan celebration of fertility. Again, a calendar was deemed the most appropriate place for a message about the nation's future.

In January 1901 *Punch* carried a large pull-out illustration, an image of the new century, called 'The Dawn' (Figure 18). This also has echoes of Delacroix, with the new century striding out in a very similar pose to Liberty, this time with the staff of 'science' in one hand and the other hand raised to shield her eyes so that she can look forward into the increasing light of the future. Here Father Time, rather than a revolutionary soldier, ushers her forward, and at her feet lie not bodies but volumes of knowledge on which her faith in the future is grounded. She has the draped toga that is a ready allusion to classical victory, but instead of the fertile revelation of the slipped chiton, the toga is embellished over one breast with a large solar image, drawing attention to her youth and fertility without affronting Victorian sensibilities. If she is the rising sun that brings the dawn (and the necklace announcing the twentieth century suggests that she is), it is her breast that bears the solar warmth. The roses that are a reference to her virginal status lie around her feet, while the briar trailing on the ground suggests the impossibility of halting her progress. The book of the future, whose pages are being turned by Mr Punch's little dog, lies open but blank, ready for great achievements to be inscribed.

Although the young calendar girls of *Punch*'s illustrations herald a development which was to go from strength to strength, the older, matronly image of Britannia also had a role to play in constructing

Figure 18: 'The Dawn', *Punch* (1901), Battye Library
This later drawing from Linley Sambourne (who produced the illustration for August 1884) seems strongly influenced by Delacroix's *Liberty Guiding the People*. The twentieth century strides out confidently with the staff of science in her hand.

images of the future. Figure 19 shows Britannia sending new year's greetings to the colonies, reminding them that the new year was about to begin (although many would not have subscribed to the Gregorian calendar). This is not simply a celebration of a holiday; it is Britannia as mother of the family of Empire, her warrior's trident put aside, organising her children to be temporally efficient. Anne McClintock has noted how 'Britain's national identity takes imperial form' in the use of images of Britannia in advertising, 'converting the narrative of imperial Progress into mass-produced *consumer spectacle*'.[24] Woman stood not only for the nation but for its pedagogical role, in this case teaching the colonies to plan for the future.

Imperial imagery figures prominently in calendars. Throughout the nineteenth century, calendar illustrations featured slaves with their shackles being cut due to British endeavours, or natives looking to Victoria for prosperity and freedom. By 1900 the young débutante had the world at her feet. In 1901 Father Time rested his foot on a globe. Such images demonstrate how middle-class faith in the future was built on the promise of economic growth, which was in turn dependent on imperial expansion. Calendars, then, were used to disseminate a sense of a bright future based on Empire.

However, this depiction included a representation of the 'foreign' as inferior. Even Mr Punch himself might find his Italian origins placing him at a disadvantage. In the 1900 image his florid face contrasts unfavourably with the girl's whiteness. Débutantes were supposed to wear as much white as they could, perhaps in multiple layers of tulle, with white flowers and feathers in their hair, but Miss 1900's status is conferred not only by her clothing but also by an extraordinary whiteness of skin. These two, the British new year and Mr Punch, may be in collusion as they step out into the new world, but she is clearly his superior, in national identity, youthful courage, and not least in race.

In the same year as this *Punch* cartoon by Bernard Partridge, a literary depiction of imperialism was published that carried very similar themes. Joseph Conrad's *Lord Jim*, published in 1900, explores the life of 'a raw youth' who dreams of heroism. In the final pages of the novel, the hero's actions in building a system of defence around himself in his tropical hideaway are described approvingly by Marlow, who is telling the story. The ditch, the earth wall, the guns, are all evidence of Jim's 'judicious foresight, his faith in the future'.[25] Yet we know that this approval is laced with irony, since if Jim had

PUNCH, OR THE LONDON CHARIVARI.—January 4, 1899.

A NEW YEAR'S GREETING.

Figure 19: 'A New Year's Greeting', *Punch* (1899), Battye Library
A hope for peace is evident in the message taken by these dove-like emissaries of a Britain which was, in fact, about to become embroiled in a costly and controversial South African war. The sky towards the horizon is full of wings: testimony to the vast geographical reach of the Empire. Britannia has set aside her warlike trident, and is here the well-organised mother, sending out greetings to remind the colonies of their place not only within British space but also within British time.

survived there would be no need for Marlow's narrative; and in any case, the events being described are long since over by the time the reader is made aware of them, since they are being narrated while a packet with Jim's last note in it is being opened: Jim's hope in the future actually belongs to the past.

Faith in the future, then, does Jim no good. Yet, like so much else about the dénouement of his life, it marks him out as 'one of us', a term which Conrad uses repeatedly to signify a British gentleman. It is a mark of civilisation, Conrad suggests, to believe in some form of progress, the future. Only a writer with doubts about imperialism, like Conrad, could complicate an example of this belief with a patent demonstration of its failure.

The danger in Jim's situation is that he has stepped outside the boundaries. These are revealed as being not only about place, but also about time. The foreboding surrounding the account of his final days is increased by clues that he is outside a Western sense of temporal measurement. The final note he pens before his death contains no date, a feature that Marlow acknowledges as significant: 'What is a number and a name to a day of days?'[26] Is it Jim's position outside the naming and numbering of days that dooms his faith in the future? As the packet containing his last note is opened in a London flat, presumably by Conrad himself, the reader's eye is drawn to a vista beyond the confines of the room:

> The spires of churches, numerous, scattered haphazard, uprose like beacons on a maze of shoals without a channel; the driving rain mingled with the falling dusk of a winter's evening; and the booming of a big clock on a tower, striking the hour, rolled past in voluminous, austere bursts of sound, with a shrill vibrating cry at the core ...
>
> No more horizons as boundless as hope, no more twilights within the forests as solemn as temples ... The hour was striking! No more! No more![27]

As Conrad turns to the packet and immerses himself in another place, another time, he draws the heavy curtains, shutting out the English city and its clock tower. The sound of the hour must be excluded before the reader can journey to the timelessness of the East Indies.

The gulf that separates the two worlds, civilised London and the forests of Patusan, is both spatial and temporal. The powerful sound of the booming clock – no doubt that great symbol of British power, Big Ben – speaks in an imperative which brings Jim's adventure to an end ('No more! No more!'). While Jim has been in Patusan, he has been, as it were, timeless, or at least outside Western time. As he falls from power and happiness, the image of the London clock brings the reader back to reality, back to a temporal rationality.

As a sailor, it is perhaps not surprising that Conrad should have used the imagery of time. The accurate observation of celestial time is a prerequisite for maritime safety, and the routine of four hourly watches may have been one of the earliest examples of universal measurement. Yet far from eulogising temporal efficiency, Conrad's description of the city clock among the 'scattered haphazard' spires of churches invites unfavourable comparison with 'twilights within the forests as solemn as temples'. The inescapable ascendancy of the former cannot be questioned, yet it contains within it the acknowledgement of a cost: the 'shrill vibrating cry at the core'. The naming and numbering of days will no doubt come to those who are forced to accept British rule. The power of 'austere' time is inexorable, as Conrad had already shown in *The Secret Agent*, where his ill-fated revolutionary was defeated by the steep incline leading up to the Greenwich-meridian clock. Yet Conrad wonders whether there might not be something 'mightier than the laws of order and progress'.

Perhaps whether or not the future will actually be happy is not the point. Jim, after all, achieves a future greater than happiness: his honour and peace of mind are rescued from the agonies of his previous sense of failure. We must believe in the future in order to be 'one of us'. Following Conrad, we might conclude that it is loss of faith in the future which is one of the distinguishing marks of a postmodern Britain, where the international calendar girls of globalised culture no longer signify national destiny and progress.

Conclusion

This study has focussed chiefly on nineteenth-century ideas about time that attracted the label 'superstition': astrology, fortune-telling, and prophetic dreaming. The premise has been that if we put aside the question of whether these ideas were right or wrong, the historical context of such beliefs may reveal fresh insights about the rise of individualism and its contribution to the intellectual high ground taken by 'Western' science. Perhaps surprisingly, most avenues of enquiry have led to answers in which gender and sexuality have been important issues. The crusaders against 'wrong' knowledge have turned out to be not the scientist and statistician, upholders of rationalism, but rather the fervent Christian, fearing the destabilising effect that unorthodox beliefs might have on morality. It is hard to escape the conclusion that the nineteenth century's campaigns against superstition were not only to do with education and enlightenment but also with the increasingly circumscribed roles of women, and fears that traditional beliefs about prediction endangered the respectable boundaries of sexual behaviour.

There is now a very large literature exploring the relationship between gender and knowledge. Most of this owes a great deal to Foucauldian insights about the way in which meaning is produced. Where the social construction of knowledge is a postulate, gender can be shown to be a factor influencing what knowledge is validated, and how that process is accomplished. Within the history of science, writers such as Donna Haraway, Mary Poovey, and Sandra Harding have treated knowledge not as the outcome of a search for universal, transcendental truth, but as located in specific historical contexts.[1] In *A History of the Modern Fact* Mary Poovey explains the role of historical epistemology in such analyses:

> [H]istorical epistemology assumes that the categories by which knowledge is organized ... inform what can be known at any given time, as well as how this knowledge can be used ... Insofar as historical epistemology assumes that the categories by which knowledge is organized change over time, it is less a study of the inexorable march of 'science' toward a fully adequate description

of nature than an investigation of those developments that have increasingly made Westerners believe this march is under way.[2]

The march of science was once thought to be the main reason for the disappearance of superstitious beliefs. If we take Mary Poovey's lead and examine the historical context of the late eighteenth and nineteenth centuries in order to discover more about why it was thought that superstition had been defeated, we find society in Western Europe in general and in Britain in particular overshadowed by the enormous edifice of imperialism. The imperial experience permeated the domestic world. Perhaps the most influential study of this process has been Edward Said's analysis of the way in which the colonised 'other' became a fundamental part of the formation of the knowledge on which modern Western society was built. Importantly for this study of superstition, the term 'orientalism', which Said coined in order to designate the use of the East as the repository of 'otherness' in Western thought, can also be applied to the marginalisation of unrespectable popular beliefs. For example, the emphasis on gypsies as purveyors of fortune-telling in Britain is actually an 'orientalist' trope. When legislation against vagrancy referred to the 'Egyptian offence', that is, fortune-telling, it was expressing the generally accepted opinion that the gypsies were travellers who had come from 'the orient', bringing with them strange Eastern beliefs in magic and prediction.

When the 'East' was not invoked as the source of pernicious superstition, other countries such as Germany or France were blamed. Even astrology was often called the 'Chaldean' art. Mistaken beliefs, especially the morally dangerous ones, were not considered as springing direct from British culture, but rather were thought to be imported from abroad.

Building on Said's work, Homi Bhabha has added to post-colonial theory an emphasis on ambivalence, suggesting that colonists were both attracted to and repelled by the 'native' and the 'exotic'.[3] Again, this analysis fits domestic society's dealings with gypsies and fortune-tellers, as earlier chapters have shown. The 'foreign' practices which the English press attributed to gypsies, such as fraud and theft, might be a nuisance for the ordinary householder, but much more worrying was the skill which gypsies seemed to have of enticing women to stray morally. The attraction of the magician was dangerous. Many members of the middle classes were drawn to gypsy encampments, we are told, particularly women. Reforming literature was aimed not at the gypsy or fortune-teller but at their potential 'victim', the woman who was likely to be deceived and corrupted. Although legislation did not explicitly refer to sexual danger, this was an important part of the gypsy stereotype. When Samuel Bamford came across a gypsy camp for the first time in his travels across England, he was fascinated with a 'superb being' of great beauty, whose raven hair 'fell in graceful locks over her shoulders and below her bosom', and whose features he had seen only 'on the statues of oriental nymphs and goddesses of antiquity'.[4] The 'other' is dark, tempting, desirable and dangerous. Even when the fortune-teller, like Tawny Rachel, is not specifically described as a gypsy, the suspicion of foreign sexuality clings. Rachel's name is Jewish, her nickname, 'tawny', means 'dark', and she travels the country. Although the county of Somerset may be glad to be rid of her, there is never any suggestion that this is where she 'really' comes from, only that it is the scene of her current travels. Her husband, poacher Giles, is blessed with a solid Anglo-Saxon name, but this may be all the more reason to be suspicious of the pair. Such a blending of names hints further at the sexual danger of gypsy beauty, if a travelling woman like Rachel can ensnare a local man.

Fortune-telling, then, was portrayed as 'foreign'. As this study has shown, several associated aspects of unorthodox belief about time were also marginalised as 'other', particularly a sense of pessimism about the future. Responsible time management was presented as important to a sense of national identity, and within the home it could be seen as a valuable contribution to imperial destiny. The division of temporal practices into 'scientific' and 'superstitious' reflects an increasing desire to demarcate a public sphere in which both memory and planning could be evaluated according to generally accepted criteria:

> At the heart of the attack on superstitious modes of thinking and behaving was the desire to generate a far more precise and disciplined attitude towards time. The minutes and hours of the day were to be saved and used to greater advantage, and the cycles of the year were to be placed in the context of a more ordered, formal and public notation of history.[5]

This marking out of time as the stable material from which national narratives of both past and future might be put together has never been entirely successful. The belief that some individuals are able to see (or in some other way sense) the future is a resilient popular belief, resurfacing in nineteenth-century mesmerism and spiritualism, as well as in twentieth-century charismatic revivals and magazine astrology. The case studies presented here help to explain why such beliefs, however widely held, are consistently relegated to the unrespectable.

Johannes Fabian stresses the cross-cultural importance of understanding the connection between magic and beliefs about time, and the political nature of this connection, taking politics in its widest sense as being about power relations. He points to the assumptions that anthropologists are prone to make when they deny 'that all temporal relations ... are embedded in culturally organized praxis'. Anthropologists, he writes, have little difficulty admitting this axiom as long as it is predicated on a specific culture, 'usually one that is not their own', but they have more difficulty putting themselves into the loop:

> To cite but two examples, relationships between the living and the dead, or relationships between the agent and object of magic operations, presuppose cultural conceptions of contemporaneity. To a large extent, Western rational disbelief in the presence of ancestors and the efficacy of magic rest on the rejection of ideas of temporal coexistence implied in these ideas and practices. So much is obvious. It is less clear that in order to study and understand ancestor cult and magic we need to establish relations of coevalness with the cultures that are studied. In that form, coevalness becomes the ultimate assault on the protective walls of cultural relativism. To put it bluntly, there is an internal connection (one of logical equivalence and of practical necessity) between ancestor cult or magic and anthropological research qua conceptualizations of shared Time or coevalness. Paraphrasing an observation by Owusu, I am tempted to say that the Western anthropologist must be haunted by the African's 'capricious ancestors' as much as the African anthropologist is 'daunted' by 'Malinowski, Evans-Pritchard, Fortes, Mair, Gluckman, Forde, Kabbery [sic], Turner, Schapera, and the Wilsons, among others'.[6]

In other words, the 'Western' anthropologist, and indeed any 'Westerner' in contact with ancestor belief, must decide either (a) that it is true, (b) that it is false, or be driven to the far-reaching conclusion (c) that there is more to knowledge than the dualistic division between truth and error.

Bearing in mind, then, the persistence of magical belief, even in Western culture, what is our sense of time, and how have we come by it? Surely a major reason for considering this question is in order to be able to communicate with those whose time sense may differ. If, indeed, we believe that the nineteenth and twentieth-century European condemnation of prediction is a universal and enduring truth, then we must consider what we make of the many non-Western cultures in which some form of 'prognostication' is an acceptable form of knowledge. If, however, we see the condemnation of fortune-telling as being much more a political act, then we are open to see the construction of scientific certainties as 'local knowledges' whose universal claims must each be negotiated. The purpose of such a debate, already strenuously underway across many forms of knowledge, is to explore our global similarities and differences, to understand what makes us both 'uniquely individual' and at the same time 'commonly human'.[7] It is my hope that a close investigation of the marginalisation of superstitious beliefs may contribute to such a debate.

Notes

Introduction

1. Tim Hilton, 'Preface', in Samuel Bamford, *Passages in the Life of a Radical* (London: Macgibbon & Kee, 1967; 1st pub. 1844), p. 6, and Bamford, *Passages*, pp. 100, 292. The attempt to protect against witchcraft mentioned by Bamford was not an isolated case. In a study of the archaeology and history of folk magic in Britain, Brian Hoggard has found evidence of dried cats, horse skulls, shoes, witch-bottles and charms, all concealed for counter-witchcraft purposes, well into the nineteenth and twentieth centuries. Details about his research can be found on <www.folkmagic.co.uk>.

2. Ibid., p. 108. Of the three young men involved in the experiment, one is found wandering out of his mind, one falls mysteriously ill and dies, and the third, the least implicated in the magic, escapes to tell the story, but, significantly, meets Samuel Bamford in prison. He, too, later met 'a mysterious fate', disappearing, leaving behind a 'young and handsome' wife.

3. Alan Dawe, 'Theories of Social Action', in Tom Bottomore and Robert Nisbet, eds, *A History of Sociological Analysis* (London: Heinemann, 1978), pp. 362–417.

4. A. Quételet, 'Sur la possibilité de mesurer l'influence des causes qui modifient les éléments sociaux' (letter to M. Villermé), in *Correspondances mathématiques et physiques* (1832), vol. 7, p. 344, cited in Ian Hacking, 'Nineteenth Century Cracks in the Concept of Determinism', *Journal of the History of Ideas* (1983), pp. 455–75.

5. William Farr, *The Fourth Session of the International Statistical Congress*, cited in ibid., p. 472.

6. The relationship of knowledge and power is famously central to post-structural theory, as posited by Michel Foucault, in, for example, *Discipline and Punish: the Birth of the Prison* (New York: Pantheon Books, 1977; 1st pub. Paris: Gallimard, 1966, trans. Alan Sheridan). The cross-cultural implications of this were expounded by Edward Said in the highly influential *Orientalism* (New York: Pantheon Books, 1978). However, it is worth remembering that the knowledge–power nexus has been examined since antiquity.

7. Mary R. O'Neil, 'Superstition', in Mircia Eliade, ed. in chief, *Encyclopaedia of Religion*, vol. 14 (New York: Macmillan, 1987), p. 165.

8. Max Weber, *The Protestant Ethic and the Spirit of Capitalism* (London: Unwin, 1930), cited in Ken Morrison, *Marx, Durkheim, Weber: Formations of Modern Social Thought* (London: Sage, 1995), p. 218.

9. This term was used by Anthony Giddens in *Modernity and Self-Identity: Self and Society in the Late Modern Age* (Cambridge: Polity, 1991), p. 111,

and appears in Ziauddin Sardar, ed., *Rescuing all our Futures: the Future of Futures Studies* (Westport, Conn.: Praeger, 1999).

10. Owen Davies, *Witchcraft, Magic and Culture 1736–1951* (Manchester: Manchester University Press, 1999), p. 107.

11. Kate Purcell, *More in Hope than Anticipation: Fatalism and Fortune Telling amongst Women Factory Workers* (Manchester: University of Manchester Sociology Department, Studies in Sexual Politics, no. 20, 1988).

12. Sardar, *Rescuing all our Futures*, p. 2.

13. Ibid., p. 28.

14. David Vincent, *Literacy and Popular Culture: England 1750–1914* (Cambridge: Cambridge University Press, 1989), p. 157: 'Modern communications, principally the printed word but also the associated forces of new transport systems and gas and electricity, were literally and metaphorically lightening the darkness of those hitherto isolated from their influence'; Mark Smith, *Religion in Industrial Society*: *Oldham and Saddleworth, 1740–1865* (Oxford: Oxford University Press, 1994), pp. 262–4. The impression that 'traditional' beliefs failed to survive in an urban setting may perhaps be reinforced by the influential work that has been done on their persistence in rural settings, e.g. James Obelkevich, *Religion and Rural Society: South Lindsey, 1825–1875* (Oxford: Clarendon Press, 1976); Pamela Horn, *The Rural World 1780–1850: Social Change in the English Countryside* (New York: St Martin's Press, 1980); Bob Bushaway, *By Rite: Custom, Ceremony and Community in England 1700–1880* (London: Junction, 1982); Bob Bushaway, '"Tacit, Unsuspected, but still Implicit Faith": Alternative Belief in Nineteenth-Century Rural England', in Tim Harris, ed., *Popular Culture in England, c. 1500–1850* (New York: St Martin's Press, 1995), pp. 189–215.

15. Keith Thomas, *Religion and the Decline of Magic* (Harmondsworth: Penguin, 1985, 1st pub. London: Weidenfeld & Nicolson, 1971), ch. 22, p. 794. In contrast, however, see Callum G. Brown, 'Did Urbanization Secularize Britain?', *Urban History Yearbook* (1988), pp. 1–14.

16. Johannes Fabian, *Time and the Other: How Anthropology Makes its Object* (New York: Columbia University Press, 1983).

17. Anthony F. Aveni, *Empires of Time: Calendars, Clocks, and Cultures* (New York: Basic Books, 1989); Bonnie Urciuoli, 'Time, Talk, and Class: New York Puerto Ricans as Temporal and Linguistic Others', in Henry J. Rutz, *The Politics of Time* (Washington DC: American Ethnological Society Monograph Series, No. 4, 1992), pp. 108–26.

18. Carol J. Greenhouse, *A Moment's Notice: Time Politics across Cultures* (Ithaca, NY: Cornell University Press, 1996).

19. Kenneth Bock, 'Theories of Progress, Development, Evolution', in Tom Bottomore and Robert Nisbet, eds, *A History of Sociological Analysis* (London: Heinemann, 1979), pp. 39–79.

20. Lori Anne Loeb, *Consuming Angels: Advertising and Victorian Women* (New York: Oxford University Press, 1994), p. 47.

21. J. B. Bury, *The Idea of Progress: an Inquiry into its Origin and Growth* (London: Macmillan, 1924; 1st pub. 1920), pp. 334–5.

22. McKenzie Wark, 'Spooked by postmodernism', *The Australian*, 18 September 1999, p. 42.

23. Arthur O. Lovejoy and George Boas, eds, *Primitivism and Related Ideas in Antiquity* (Baltimore, 1935), cited in W. Warren Wagar, 'Modern Views of the Origins of the Idea of Progress', *Journal of the History of Ideas*, vol. 28 (January–March 1967), p. 55.
24. For example, Jean Baudrillard, 'Hystericizing the Millennium', <http://www.ctheory.com/a-hystericizing_the.html>, originally published as *L'Illusion de la fin: ou La grève des événements* (Paris: Galilée, 1992; trans. Charles Dudas).
25. I use the term 'social progress' here to distinguish it from scientific concepts associated, for example, with evolutionary biology, which are currently influential in scientific epistemology.
26. David Spadafora, *The Idea of Progress in Eighteenth-Century Britain* (New Haven, Conn.: Yale University Press, 1990), p. 6. The literature on progress can also be approached through Reinhart Koselleck, *Futures Past: on the Semantics of Historical Time* (Frankfurt am Main, 1979; trans. Keith Tribe, Cambridge, Mass: MIT Press, 1985). A survey of the large literature on this subject can be found in Wagar, 'Modern Views', pp. 59–69.
27. Wagar, 'Modern Views', p. 56.
28. Spadafora, *Progress*, pp. 6, 385–6.
29. Peter Burke, *Popular Culture in Early Modern Europe* (London: T. Smith, 1978).
30. The description is from Graeme Davison, *The Unforgiving Minute: How Australia Learned to Tell the Time* (Melbourne: Oxford University Press, 1993).
31. Samuel L. Macey, *Clocks and the Cosmos: Time in Western Life and Thought* (Hamden, Conn.: Archon Books, 1980), esp. pt II, pp. 65–120.
32. Mortimer Collins, *Thoughts in My Garden* (ed. Edmund Yates, 2 vols, London, 1880), vol. 1, pp. 6–8. Comments on 'telling the bees' can also be found in Obelkevich, *Religion and Rural Society*, pp. 296–9. Several references to clocks stopping with the death of their owner can be found in nineteenth-century fiction, such as *Master Humphrey's Clock*, the weekly magazine produced by Charles Dickens, in the final edition of which Master Humphrey dies and the clock stops for ever. See Macey, *Clocks and the Cosmos*, p. 189.
33. Ann Curthoys, *AHA Bulletin*, December 1996, referring to Heather Goodall, *From Invasion to Embassy: Land in Aboriginal Politics in New South Wales, 1770–1972* (St Leonards, NSW: Allen & Unwin, 1996).
34. Dipesh Chakrabarty, 'Minority Histories, Subaltern Pasts', *Perspectives*, vol. 35, no. 8 (November 1997).
35. Anne McClintock, *Imperial Leather: Race, Gender and Sexuality in the Colonial Contest* (New York: Routledge, 1995), p. 10.
36. The editors of *The Post-Colonial Studies Reader* suggest that limiting post-coloniality to 'after' colonialism misses the 'continuing process of resistance and reconstruction'. Bill Ashcroft, Gareth Griffiths and Helen Tiffin, eds, *The Post-Colonial Studies Reader* (London: Routledge, 1995), p. 2.
37. Anthony Giddens, *The Consequences of Modernity* (Cambridge: Polity Press in association with Basil Blackwell, 1990).

38. Joseph E. McGrath, *The Social Psychology of Time: New Perspectives* (Newbury Park: Sage, 1988), p. 45.
39. For a more detailed, theoretical discussion of the cross-cultural implications of 'Western' science, see Sandra Harding, 'Is Modern Science an Ethnoscience? Rethinking Epistemological Assumptions', in Emmanuel Chukwudi Eze, *Postcolonial African Philosophy: a Critical Reader* (Cambridge, Mass.: Blackwell, 1997), pp. 45–70. Harding's endnote 4, pp. 66–7, provides a useful survey of literature on this subject.
40. Robert Poole, *Time's Alteration: Calendar Reform in Early Modern England* (London: UCL Press, 1998), p. 164.
41. Ibid., p. 178.
42. E. P. Thompson, 'Time, Work-Discipline and Industrial Capitalism', *Past and Present*, vol. 38 (1967), pp. 56–97.
43. Daniel Goleman, *Emotional Intelligence* (New York: Bantam Books, 1995).

Chapter 1

1. The *Oxford English Dictionary* defines the meridian as 'that great circle of the celestial sphere which passes through the celestial poles and the zenith of any place on the earth's surface'. Although it is often taken to refer to the highest point, in reference to the sun's position at midday, it can also refer to the centre point between the apparent halves created by this position, hence its use as 'the middle period of a man's life'.
2. Discussions leading to the choice of Greenwich are outlined in Derek Howse, *Greenwich Time and the Discovery of the Longitude* (Oxford: Oxford University Press, 1980), pp. 138–51. One of the main considerations was that the prime meridian should pass through an astronomical observatory 'of the first order'. This really amounted to Greenwich, Paris, Berlin, or Washington. It was also stated at the earlier Rome conference that 90 per cent of navigators engaged in foreign trade already calculated their longitudes from Greenwich: Howse, *Greenwich Time*, p. 137.
3. Philip S. Bagwell, *The Transport Revolution from 1770* (London: Batsford, 1974).
4. Ibid., p. 49.
5. Ibid., p. 53.
6. Howse, *Greenwich Time*, p. 87.
7. *Punch*, 26 August 1865, p. 74; 29 July 1865, p. 33.
8. *Punch*, 5 August 1865, p. 44.
9. Joseph Conrad, *The Secret Agent: A Simple Tale* (London: Penguin, 1988; 1st pub. 1907), pp. 67–8. The significance of Greenwich in spatial terms, as the home of the dividing line between eastern and western hemispheres of the globe, is touched on in Robert J. C. Young, *Colonial Desire: Hybridity in Theory, Culture and Race* (London: Routledge, 1995), p. 1.
10. This continued until 1911. Alun C. Davies, 'Greenwich and Standard Time', *History Today*, vol. 28 (1978), pp. 194–9.
11. Dava Sobel, *Longitude: the True Story of a Lone Genius who Solved the Greatest Scientific Problem of his Time* (New York: Walker, 1995).

12. Howse, *Greenwich Time*, p. 67.

13. Ibid., p. 99.

14. John Darwin, *The Triumphs of Big Ben* (London: Robert Hale, 1986). Darwin discusses the wider use of the name 'Big Ben', originally meant only for the bell, but now used for the clock, the clock tower, and the locality (pp. 14–15).

15. John M. MacKenzie, '"In Touch with the Infinite": the BBC and the Empire, 1923–53', in John M. MacKenzie, ed., *Imperialism and Popular Culture* (Manchester: Manchester University Press, 1986), pp. 165–91.

16. 43 & 44 Vict., c. 9, cited in Howse, *Greenwich Time*, pp. 114–15.

17. Ibid.

18. Robert Poole, *Time's Alteration: Calendar Reform in Early Modern England* (London: UCL Press, 1998), p. 167.

19. Maureen Perkins, *Visions of the Future: Almanacs, Time and Cultural Change 1775–1870* (Oxford: Oxford University Press, 1996). A detailed discussion of the differences between the terms 'almanac' and 'calendar' is given on pp. 14–16.

20. C. G. Harding, ed., *The Republican*, vol. 1 (London, 1848), p. 36. Republished in Dorothy Thompson, ed., *Chartism: Working-Class Politics in the Industrial Revolution* (New York: Garland Publishing, 1986).

21. Western Australian Archives, 2898/A, item 157, in file 586/2.

22. Richard Jefferies, *Wild Life in a Southern County* (1st pub. London, 1879; reprinted Nelson, n.d.), p. 110, quoted in Bob Bushaway, *By Rite: Custom, Ceremony and Community in England 1700–1880* (London: Junction, 1982), p. 48.

23. Poole, *Time's Alteration*, p. 21.

24. *Hansard*, vol. 151 (28 June 1858), p. 479.

25. 'Preliminary Observations', *British Almanac* (1828), quoted in the *Athenaeum* (2 January 1828), p. 5. The plot referred to was the Jacobite plot to assassinate William III, for which Sir John Friend was executed.

26. Royal birthdays had appeared in *Wing's* before, but as part of a separate chronology, not in the main body of the calendar. The practice of noting ecclesiastical holy days in red in both prayer book and calendar was the origin of the modern term 'red-letter day' for a happy occasion.

27. Ronald Hutton, *The Stations of the Sun: a History of the Ritual Year in Britain* (Oxford: Oxford University Press, 1996), p. 112.

28. John K. Walton and Robert Poole, 'The Lancashire Wakes in the Nineteenth Century', in Robert D. Storch, ed., *Popular Culture and Custom in Nineteenth-Century England* (London: Croom Helm, 1982), pp. 100–24.

29. Douglas A. Reid, 'The Decline of Saint Monday, 1766–1876', *Past and Present*, vol. 71 (May 1976), pp. 76–101.

30. J. A. R. Pimlott, *The Englishman's Holiday: a Social History* (London: Faber & Faber, 1947), p. 80.

31. Ibid., p. 81.

32. Poole, *Time's Alteration*, p. 21.

33. *Gentleman's Magazine*, vol. 72, part 2 (1802), pp. 1193–5.

34. P. W. Wilson, *The Romance of the Calendar* (New York: Norton, 1937).

35. Meredith N. Stiles, *The World's Work and the Calendar* (Boston: The Gorham Press, 1933).

36. *Vo-Key's Royal Pocket Index, Key to Universal Time Upon the Face of Every Watch and Clock* (London, 1901), BL 8560.a.47.
37. Anthony Letts, 'A History of Letts', in *Letts Keep a Diary*, catalogue to the 'Exhibition of the History of Diary Keeping in Great Britain from the Sixteenth to Twentieth Century in Commemoration of 175 years of Diary Publishing by Letts, The Mall Galleries, London, 28 September–25 October 1987' (London: Charles Letts, 1987), pp. 29–31.
38. David Vincent, *Literacy and Popular Culture: England 1750–1914* (Cambridge: Cambridge University Press, 1989), p. 192.
39. *Punch*, 12 August 1865, p. 54.
40. Loeb, *Consuming Angels*, p. 8, n. 20.
41. Anthony Giddens links insurance with the 'colonisation of the future': 'Insurance ... is part of a more general phenomenon concerned with the control of time which I shall term the *colonisation of the future.*' Giddens, *Modernity and Self-Identity: Self and Society in the Late Modern Age* (Cambridge: Polity, 1991), p. 111.
42. Cocoatina, 'Black and White', 9 April 1898, p. 507, cited in Loeb, *Consuming Angels*, p. 56.
43. *British Almanac* (1901), p. x.
44. *The Times* (3 December 1844), p. 1, col. 5.
45. *British Almanac*, 1901, p. vii.
46. Wilkie Collins, *The Moonstone* (Harmondsworth: Penguin, 1955; 1st pub. 1868), p. 23.
47. Stephen Kern, *The Culture of Time and Space 1880–1918* (Cambridge, Mass.: Harvard University Press, 1983), pp. 115–17.
48. Royal Stationers' Company, Box B, Envelope 9, 'Supplementary Material', letter from C. Duncan Cross on 22 March 1909 to the General Manager of Cassell & Co. for a report to the Stationers.
49. Cited in Howse, *Greenwich Time*, p. 102.
50. Walter E. Houghton, *The Victorian Frame of Mind 1830–1870* (New Haven and London: Yale University Press, 1957), p. 7.
51. Jeremy Bentham, *A Fragment on Government* (1776), in *Works*, vol. 1: 227: p. 1 ('Preface to the First Edition'), cited in Spadafora, *The Idea of Progress in Eighteenth-Century Britain* (New Haven, Conn.: Yale University Press, 1990), p. 50.
52. Kern, *Time and Space*, p. 279.
53. A. J. P. Taylor, *War by Time-Table: How the First World War Began* (New York, 1969), p. 7, cited in Kern, *Time and Space*, p. 279.

Chapter 2

1. A fuller version of this episode is told in Maureen Perkins, *Visions of the Future: Almanacs, Time and Cultural Change 1775–1870* (Oxford: Oxford University Press, 1996), ch. 6.
2. Morgan Brierley, 'Weather Forecasts and Prophecies', *Notes and Queries*, 11 January 1879, pp. 10–13.
3. Fortune-telling had been liable to prosecution earlier than this, though not in association with witchcraft. Keith Thomas, *Religion and the Decline*

of Magic (Harmondsworth: Penguin, 1985, 1st pub. London: Weidenfeld & Nicolson, 1971), p. 292.

4. 5 Geo. IV, c. 83. The first offences in a much longer list are: '[E]very person pretending or professing to tell Fortunes, or using any subtle Craft, Means or Device, by Palmistry or otherwise, to deceive and impose on any of His Majesty's Subjects; every Person wandering abroad and lodging in any Barn or Outhouse, or in any deserted or unoccupied Building, or in the open Air, or under a Tent, or in any Cart or Waggon, not having any visible Means of Subsistence, and not giving a Good Account of himself or herself; every Person wilfully exposing to view, in any Street, Road, Highway or public Place, any obscene Print, Picture or other indecent Exhibition; every Person wilfully, openly, lewdly and obscenely exposing his Person in any Street, Road or public Highway, or in the View thereof, or in any Place of public Resort, with Intent to insult any Female; every Person wandering abroad and endeavouring by the Exposure of Wounds or Deformities to obtain or gather Alms.'

5. Thomas, *Religion and the Decline of Magic*, esp. ch. 22.

6. Ibid., p. 794.

7. *Statutes at Large*, 17 Geo. II, c. 5.

8. Hugh Cunningham, *Leisure in the Industrial Revolution c.1780–c.1880* (London: Croom Helm, 1980), esp. pp. 12–13, 192–9.

9. See under 'diversion', in Samuel Johnson, *A Dictionary of the English Language* (2 vols; London, 1755), cited by Robert Malcolmson, *Popular Recreations in English Society 1700–1850* (Cambridge: Cambridge University Press, 1973), p. 4.

10. Samuel Bamford, *The Autobiography of Samuel Bamford*, vol. 1: *Early Days*, ed. W. H. Chaloner (London: Frank Cass, 1967; 1st pub. 1849), pp. 202–4.

11. *The Times*, 12 October 1807, p. 3.

12. Maurice J. Quinlan, *Victorian Prelude: a History of English Manners 1700–1830* (London: Frank Cass, 1965; 1st pub. New York: Columbia University Press, 1941), p. 204.

13. M. J. D. Roberts, 'The Society for the Suppression of Vice and Its Early Critics, 1802–1812', *The Historical Journal*, vol. 26, no. 1 (1983), pp. 164–7.

14. *Cobbett's Political Register*, vol. 4 (1803), pp. 528–31, cited in Roberts, 'The Suppression of Vice'.

15. The Society for the Suppression of Vice, *The Trial of Joseph Powell* (London, 1808), p. 17.

16. Owen Davies, *Witchcraft, Magic and Culture 1736–1951* (Manchester: Manchester University Press, 1999), p. 77.

17. Ibid., pp. 14–16.

18. David Vincent, *Literacy and Popular Culture: England 1750–1914* (Cambridge: Cambridge University Press, 1989), p. 180.

19. Edward J. Bristow, *Vice and Vigilance: Purity Movements in Britain since 1700* (Dublin: Gill and Macmillan, 1977), p. 3.

20. M. J. D. Roberts, 'Public and Private in Early Nineteenth-Century London: the Vagrant Act of 1822 and its Enforcement', *Social History*, vol. 13, no. 3 (1988), pp. 273–94. There seems to have been a rerun of

the same concern about public behaviour in open spaces at the beginning of the twentieth century. See Nan H. Dreher, 'The Virtuous and the Verminous: Turn-of-the-Century Moral Panics in London's Public Parks', *Albion*, vol. 29, no. 2 (Summer 1997), pp. 246–67.

21. John Adolphus, *Observations on the Vagrant Act and Some Other Statutes, and on the Powers and Duties of Justices of the Peace* (London: John Major, 1824), p. 4.
22. 'Returns of Persons committed under the Vagrant Laws, 1 January 1820 – 1 January 1824', *British Parliamentary Papers*, 1824, xix, pp. 211–338.
23. Society for the Suppression of Vice, *Examination before the Police Committee of the House of Commons*, 19 June 1817, p. 7.
24. Bristow, *Vice and Vigilance*, p. 59.
25. Patty Seleski, 'Women, Work and Cultural Change in Eighteenth- and Early Nineteenth-Century London', in Tim Harris, ed., *Popular Culture in England, c.1500–1850* (New York: St Martin's Press, 1995), p. 153.
26. Ibid., p. 149.
27. My thanks to Michael Roberts for this point.
28. The Society for the Suppression of Vice, *The Trial of Joseph Powell*, p. 26.
29. Owen Davies, *A People Bewitched: Witchcraft and Magic in Nineteenth-Century Somerset* (Trowbridge, Wiltshire: Bruton 1999), p. 93.
30. Mr Johnson, *The Female Fortune-Teller, A Comedy as it is Acted at the Theatre in Lincoln's-Inn Fields* (London, 1726), p. 22.
31. Z [i.e. Hannah More], *Tawny Rachel, or the Fortune Teller, with Some Account of Dreams, Omens and Conjurers* (London, 1797?).
32. Davies, *Witchcraft*, p. 228.
33. The Society for the Suppression of Vice, *The Trial of Joseph Powell*, p. 4.
34. John Ashton, *A History of English Lotteries* (London: Leadenhall Press, 1893), p. 298.
35. Ibid.
36. Seleski, 'Women, Work and Cultural Change', p. 150. More is said about the servant 'problem' in Chapter 3.
37. Ashton, *English Lotteries*, p. 310.
38. The unrespectability of dreams is discussed further in Chapter 3.
39. Vincent, *Literacy and Popular Culture*, p. 194.
40. Ibid.
41. Ross McKibbin, 'Working-Class Gambling in Britain, 1880–1939', in Ross McKibbin, *The Ideologies of Class: Social Relations in Britain 1880–1950* (Oxford: Oxford University Press, 1990), pp. 101–39, cited in Gary Cross, *Time and Money: the Making of Consumer Culture* (London: Routledge, 1993), p. 147.
42. Perkins, *Visions of the Future*, p. 70; Mark Smith, *Religion in Industrial Society: Oldham and Saddleworth 1740–1865* (Oxford: Oxford University Press, 1994), pp. 266–7.
43. This gave rise to Jonathan Swift's famous practical joke, in which he used the false persona of Isaac Bickerstaff, astrologer, to forecast Partridge's own death. See Patrick Curry, *Prophecy and Power: Astrology in Early Modern England* (Cambridge: Polity, 1989), pp. 89–91.
44. James Obelkevich, *Religion and Rural Society: South Lindsey 1825–1875* (Oxford: Clarendon Press, 1976), p. 289.

45. Martin J. Wiener, *Reconstructing the Criminal: Culture, Law, and Policy in England 1830–1914* (Cambridge: Cambridge University Press, 1990), p. 42, n. 107.
46. Sara Mendelson and Patricia Crawford, *Women in Early Modern England 1550–1720* (Oxford: Clarendon Press 1998), ch. 2.
47. The Society for the Suppression of Vice, *The Trial of Joseph Powell*, p. 26.
48. Bamford, *Early Days*, p. 214.
49. *Bath Chronicle*, 25 December 1851, cited in Davies, *A People Bewitched*, p. 104.
50. Janet Oppenheim, *The Other World: Spiritualism and Psychical Research in England 1850–1914* (Cambridge: Cambridge University Press, 1985), p. 24.
51. For a useful discussion of astrology see Patrick Curry, 'Astrology in Early Modern England: the Making of a Vulgar Knowledge', in Stephen Pumfrey, Paolo L. Rossi and Maurice Slawinski, eds, *Science, Culture and Popular Belief in Renaissance Europe* (Manchester: Manchester University Press, 1991), pp. 274–91; Curry, *Prophecy and Power*; and Curry, *A Confusion of Prophets: Victorian and Edwardian Astrology* (London: Collins & Brown, 1992).
52. The associated campaign by the Society for the Diffusion of Useful Knowledge, not connected with the Vice Society's work, is examined in Perkins, *Visions of the Future*.
53. Curry, *A Confusion of Prophets*, p. 66.
54. On the privatisation of religion, see James A. Beckford, *Religion and Advanced Industrial Society* (London: Unwin Hyman, 1989), pp. 76–7, 103–4.

Chapter 3

1. Tony James, *Dream, Creativity, and Madness in Nineteenth-Century France* (Oxford: Clarendon Press, 1995).
2. Ibid., p. 272.
3. *Chambers's Journal*, 4th series, vol. 53, no. 630 (22 January 1876), pp. 56–9.
4. C. Binz, *Über den Traum* (Bonn, 1878), p. 35, cited in Sigmund Freud, *The Essentials of Psycho-Analysis* (trans. James Strachey; London: Hogarth Press, 1986), p. 82.
5. Terry Castle, *The Female Thermometer: Eighteenth-Century Culture and the Invention of the Uncanny* (New York: Oxford University Press, 1995), p. 183.
6. Patrick Wolfe, 'On Being Woken Up: the Dreamtime in Anthropology and in Australian Settler Culture', *Comparative Studies in Society and History*, vol. 32 (1991), pp. 197–224.
7. Castle, *The Female Thermometer*, p. 184.
8. C. Lehec and J. Cazeneuve, *Oeuvres philosophiques de Cabanis* (Corpus général des philosophes français; PUF, 1956) xliv/I.597, cited in James, *Dreams*, p. 1.

9. This is not to deny an enormous literature on dreams, most of it written by men, but even those who acknowledged that prophecy might play a part in some dreams were usually ambivalent. For an excellent overview of different nineteenth-century opinions about dreams, see Merle Curti, 'The American Exploration of Dreams and Dreamers', *Journal of the History of Ideas*, vol. 27 (1966), pp. 391–416 (of interest not only in an American context). Of relevance to my argument here, Curti's survey of eighteenth-century American colonists' diaries leads to the following conclusion: '[T]he best known diarists merely report dreams [and] reflect on the rising fashion in Europe to discount their importance' (p. 393).
10. *Blackwood's Magazine*, vol. 48 (August 1840), pp. 194–204.
11. Frank Seafield, *The Literature and Curiosities of Dreams: A Commonplace Book or Speculations Concerning the Mystery of Dreams and Visions* (2 vols, London, 1865), vol. 1, p. 134.
12. Florence Nightingale, *Cassandra*, written 1852; pub. as Appendix in Ray Strachey, *The Cause* (Virago reprint, 1978, pp. 413–16); cited in Diana Basham, *The Trial of Woman: Feminism and the Occult Science in Victorian Literature and Society* (Basingstoke: Macmillan, 1992), p. 57.
13. *Chambers's Journal*, vol. 2, no. 62 (6 April 1833), pp. 77–8.
14. David Collins, *An Account of the English Colony in New South Wales* (2 vols, London, 1802), vol. 2, p. 106.
15. E. P. Thompson, *The Making of the English Working Class* (Harmondsworth: Penguin, 1986; 1st pub. London: Victor Gollancz, 1963), pp. 127–9, 420–8.
16. William Howitt, *Land, Labour, and Gold or Two Years in Victoria* (London: Longman, Brown, Green, Longmans & Roberts, 1855), p. 12.
17. Hugh McCrae, ed., *Georgiana's Journal: Melbourne 1841–1865* (Sydney, 1934, 1966), p. 38.
18. John Dunmore Lang, *Reminiscences of My Life and Times both in Church and State in Australia for upwards of Fifty Years* (ed. D. W. A. Baker, Melbourne: Heinemann, 1972), p. 80.
19. James Guest, 'A Free Press, and How it Became Free', in W. Hutton, *The History of Birmingham* (Birmingham, 1861), quoted in Joel Wiener, *The War of the Unstamped: the Movement to Repeal the British Newspaper Tax 1830–1836* (Ithaca, NY: Cornell University Press, 1969), p. 17.
20. P. E. Razzell and R. W. Wainwright, eds, *The Victorian Working Class: Selections from Letters to the Morning Chronicle*, pp. 175–6, cited in Paul A. Pickering, 'Chartism and the "Trade of Agitation" in Early Victorian Britain', *History*, vol. 76 (1991), pp. 221–37.
21. Charles Godfrey Leland, *Gypsy Sorcery and Fortune-Telling* (London, 1891), p. xvi.
22. David Vincent, *Literacy and Popular Culture: England 1750–1914* (Cambridge: Cambridge University Press, 1989), pp. 179–80).
23. Allan M. Galer, *Norwood and Dulwich: Past and Present, with Historical and Descriptive Notes* (London: Truslove & Shirley, 1890), p. 13.
24. *The Golden Dreamer, or Dreamer's Interpreter, Clearly Showing how All Things Past, Present and to Come may be Ascertained by means of Dreams* (Newcastle upon Tyne, n.d., but BL has catalogued it as 1850).

25. S. R. F. Price, 'The Future of Dreams: From Freud to Artemidorus', *Past and Present*, no. 113 (November 1986), pp. 3–37.
26. *The True Fortune Teller* (Edinburgh, 1850), p. 24.
27. Marilyn Butler, 'Antiquarianism (Popular)', in Iain McCalman, ed., *The Oxford Companion to the Romantic Age: British Culture 1776–1832* (Oxford: Oxford University Press, 1999), pp. 328–38.
28. Patrick Curry, *A Confusion of Prophets: Victorian and Edwardian Astrology* (London: Collins & Brown, 1992), p. 54.
29. *Smithson's Northallerton Almanack*, March 1899.
30. Vincent, *Literacy and Popular Culture*, p. 24.
31. Seafield, *Literature and Curiosities of Dreams*, vol. 1, p. 134.
32. Galer, *Norwood and Dulwich*, p. 11.
33. Social exploration as a late nineteenth-century phenomenon is discussed in Peter Keating, ed., *Into Unknown England 1866–1913: Selections from the Social Explorers* (London: Fontana, 1976).
34. *Dreams and Moles* (London, 1750), p. 23.
35. *The Dreamer's Oracle, being a Faithful Interpretation of Two Hundred Dreams* (Derby, 1830).
36. Castle, *The Female Thermometer*, 1995), p. 175.
37. Margaret Nancy Cutts, *Ministering Angels: a Study of Nineteenth-Century Evangelical Writing for Children* (Wormley, Herts.: Five Owls Press, 1979), p. 117. My thanks to Dr Sue Rickard for this reference.
38. Margaret Spufford, *Small Books and Pleasant Histories: Popular Fiction and its Readership in Seventeenth-Century England* (London: Methuen, 1981), p. 61.
39. See Maureen Perkins, *Visions of the Future: Almanacs, Time and Cultural Change 1775–1870* (Oxford: Oxford University Press, 1996), ch. 4.
40. Spufford, *Small Books*, ch. 7; Sara Mendelson and Patricia Crawford, *Women in Early Modern England 1550–1720* (Oxford: Clarendon Press, 1998), p. 111.
41. Alex Owen, *The Darkened Room: Women, Power, and Spiritualism in Late Victorian England*, (Philadelphia: University of Pennsylvania Press, 1990), p. 2.
42. Perkins, *Visions of the Future*, pp. 39–45.
43. Patty Seleski, 'Women, Work and Cultural Change in Eighteenth-and Early Nineteenth-Century London', in Tim Harris, ed., *Popular Culture in England c. 1500–1850* (New York: St Martin's Press, 1995), p. 156.
44. Henri-Jean Martin, 'The Bibliothèque Bleue: Literature for the Masses in the Ancien Régime', trans. David Gerard, in *Publishing History*, vol. 3 (1978), pp. 70–102.
45. Z [i.e. Hannah More], *Tawny Rachel, or the Fortune Teller; with Some Account of Dreams, Omens and Conjurers*, Cheap Repository for Religious and Moral Tracts, no. 17 (London, n.d., but catalogued in BL as possibly 1797), p. 15.
46. Wilkie Collins, *The Woman in White* (Oxford 1949; 1st pub. London, 1860), p. 27.
47. *Dreams and Moles*, p. 11.

48. *The New Infallible Fortune Teller, or a Just Interpretation of Dreams and Moles, to which are added, Rules to Foretell the Weather; Drawn up from the Strict Observance of nearly Half a Century* (Edinburgh, 1818), pp. 12–13.
49. *The Fortune Teller* (n.d.) [BL no: 12316 d. 57 (10)]
50. Keith Thomas, *Religion and the Decline of Magic* (Harmondsworth: Penguin, 1985; 1st pub. London: Weidenfeld & Nicholson, 1971); James Obelkevich, *Religion and Rural Society: South Lindsey 1825–1875* (Oxford: Clarendon Press, 1976); Ronald Hutton, *The Stations of the Sun: a History of the Ritual Year in Britain* (Oxford: Oxford University Press, 1996).
51. Herbert Leventhal, *In the Shadow of the Enlightenment: Occultism and Renaissance Science in Eighteenth-Century America* (New York: New York University Press, 1976), p. 58.
52. The Society for the Suppression of Vice, *The Trial of Joseph Powell* (London: 1808), p. 25.
53. *The Everlasting Circle of Fate or Expounder of Things that are to be; Being a New and Original Mode of Telling Fortunes* (Edinburgh, 1840).
54. *The Everlasting Circle of Fate*, p. 40. [my italics]
55. *The Golden Fortune-Teller* (Newcastle upon Tyne, 1840), p. 12.
56. Catherine Nash, 'Reclaiming Vision: Looking at Landscape and the Body', *Gender, Place and Culture*, vol. 3, no. 2 (1996), pp. 149–69. The last two quotes here are cited from the video installation discussed in Nash's article: Pauline Cummins, *Inis t'Oirr/Aran Dance*.
57. *OED*, s.v. 'salt'.
58. H. G. Wells, 'The So-called Science of Sociology', in *An Englishman Looks at the World* (London: Cassell, 1914), p. 205.
59. Mark Hillegas, *The Future as Nightmare: H. G. Wells and the Anti-Utopians* (New York: Oxford University Press, 1967).
60. Harold Bloom, ed., *Sigmund Freud's 'The Interpretation of Dreams'* (New York: Chelsea House, 1987), p. 1.
61. Sigmund Freud, 'On Dreams' (1901), in Freud, *The Essentials of Psycho-Analysis*, p. 83.
62. Ibid., p. 123.
63. Note the contrast with Foucault: 'The essential point of a dream is not so much what it brings back from the past but what it announces of the future ... the dream anticipates the moment of liberation. It foresees the story, rather than being just the repetition of a traumatic past.' Foucault, 'Dream, Imagination, and Existence', trans. F. Williams, in K. Hoeller (ed.), *Dream and Existence*, special issue of *Review of Existential Psychology and Psychiatry* (Seattle, 1954), cited in Vikki Bell, 'Dreaming and Time in Foucault's Philosophy', *Theory, Culture and Society: Explorations in Critical Social Science*, vol. 11, no. 2 (May 1994), pp. 151–64.

Chapter 4

1. John Foster Fraser, *Pictures from the Balkans* (London, 1906), cited in Jason Goodwin, *Lords of the Horizons: a History of the Ottoman Empire* (London: Chatto & Windus, 1998), p. 307.

2. Anthony Pagden, *European Encounters with the New World: from Renaissance to Romanticism* (New Haven: Yale University Press, 1993), p. 10.
3. Cited in Anthony Pagden, *The Fall of Natural Man: the American Indian and the Origins of Comparative Ethnology* (Cambridge: Cambridge University Press, 1982), p. 181.
4. Leela Gandhi, *Postcolonial Theory: A Critical Introduction* (St Leonards, NSW: Allen & Unwin, 1998), p. 32.
5. James Ferguson, *Astronomy Explained* (London, 1756), p. 1, cited in Robert Poole, *Time's Alteration: Calendar Reform in Early Modern England* (London: UCL Press, 1998), p. 166.
6. Charles Gilson, *In the Power of the Pygmies* (Milford, 1919), p. 276, cited in B. V. Street, *The Savage in Literature: Representations of 'Primitive' Society in English Fiction 1858–1920* (London: Routledge & Kegan Paul, 1975), p. 63.
7. Karl Marx, 'The Future Results of the British Rule in India' (1853), reprinted in Marx and Friedrich Engels, *The First Indian War of Independence 1857–1859* (Moscow, 1959), pp. 29–35, cited in Patrick Wolfe, 'History and Imperialism: A Century of Theory, from Marx to Postcolonialism', *The American Historical Review*, vol. 102, no. 2 (April 1997), pp. 388–420. The industrial world's role was, for Marx, to set India on its own course of historical development, a course that would eventually lead through capitalism to an Indian transition to socialism.
8. Carol J. Greenhouse, *A Moment's Notice: Time Politics across Cultures* (Ithaca, NY: Cornell University Press, 1996), p. 149.
9. Ibid., p. 147.
10. Keith Thomas, *Religion and the Decline of Magic* (Harmondsworth: Penguin, 1985; 1st pub. London: Weidenfeld & Nicholson, 1971), pp. 74–5, 78.
11. Ibid., p. 78.
12. Quoted in Hartmut Lehmann and Guenther Roth, eds, *Weber's Protestant Ethic: Origins, Evidence, Contexts* (New York: Cambridge University Press, 1993), p. 55.
13. Samuel Bamford, *Passages in the Life of a Radical* (London: Macgibbon & Kee, 1967; 1st pub. 1844), p. 291.
14. Ibid., p. 284.
15. Henry Mayhew, *London Labour and the London Poor: a Cyclopaedia of the Condition and Earnings of Those that Will Work, Those that Cannot Work, and Those that Will Not Work* (London: Frank Cass, 1967; 1st pub. 1851), vol. 1, pp. 2–3.
16. Donald Kenrick and Grattan Puxon, *The Destiny of Europe's Gypsies* (London: Heinemann, 1972), pp. 19–21.
17. Ibid., p. 22.
18. It is thought that gypsies may have been settled in this area of London (then woodland) as early as the seventeenth century. Gipsy Hill still survives as the name of part of the area. In 1668 Samuel Pepys entered in his diary: 'This afternoon my wife and Mercer and Deb went with Pelling to see the Gypsies at Lambeth and have their fortunes told; but what they did, I did not enquire.' Cited in Allan M. Galer, *Norwood and*

Dulwich: Past and Present, with Historical and Descriptive Notes (London: Truslove & Shirley, 1890), p. 10. Galer also lists the Prince of Wales (the future George III) and Lord Byron as visitors to the Norwood gypsies, p. 11.

19. David Mayall, *Gypsy-Travellers in Nineteenth-Century Society* (Cambridge: Cambridge University Press, 1988), p. 154.

20. Charles Godfrey Leland, *Gypsy Sorcery and Fortune-Telling* (London, 1891), p. xi.

21. Returns of Persons committed under the Vagrant Laws, 1 January 1820 – 1 January 1824, British Parliamentary Papers, 1824, xix (printed by the House of Commons, 26 May 1824), p. 221.

22. J.-P. Liégeois, *Gypsies, An Illustrated History* (trans. Tony Berrett, London: Al Saqi Books, 1986), pp. 90, 102.

23. *Statutes At Large*: 9 Geo. II, c. 5; 17 Geo, II, c. 5; 32 Geo. III, c. 45; 3 Geo IV, c. 40; 5 Geo. IV, c. 83.

24. 22 Hen. VIII, c. 10, the Egyptians Act.

25. John Clare, 'Autobiographical Fragments', in Eric Robinson, ed., *John Clare's Autobiographical Writings* (Oxford: Oxford University Press, 1983), p. 71. I have not made any alterations to Clare's spelling.

26. D. H. Lawrence, *The Virgin and the Gypsy* (London: Heinemann, 1980), pp. 19–20. 'Time's forelock' was an image that expressed the impossibility of ever capturing and stopping time, since it was only when he went past that one was aware of his presence, yet the forelock was on his forehead, and so could never be grasped.

27. Leland, *Gypsy Sorcery*, p. xi.

28. Owen Davies, *Witchcraft, Magic and Culture 1736–1951* (Manchester: Manchester University Press, 1999), p. 134.

29. John Ashton, *A History of English Lotteries* (London: Leadenhall Press, 1893) p. 343.

30. E. E. Evans-Pritchard, *The Nuer*, quoted in Anthony F. Aveni, *Empires of Time: Calendars, Clocks, and Cultures* (New York: Basic Books, 1989), pp. 167–8.

31. Cited in Donald E. Brown, *Human Universals* (Philadelphia: Temple University Press, 1991), p. 27.

32. Johannes Fabian, *Time and the Other: How Anthropology Makes its Object* (New York: Columbia University Press, 1983), p. 31.

33. *Western Australian Almanac* (Perth: 1842), pp. xii, xxvii.

34. Cited in E. J. Stormon, ed., *The Salvado Memoirs: Historical Memoirs of Australia and Particularly of the Benedictine Mission of New Norcia of the Habits and Customs of the Australian Natives* (Nedlands, WA: UWA Press, 1977), p. 110.

35. Cited in Charles Fox and Marilyn Lake, eds, *Australians at Work: Commentaries and Sources* (Fitzroy, Vic.: McPhee Gribble, 1990), p. 81.

36. W. E. H. Stanner, *White Man Got No Dreaming* (Canberra: Australian National University Press, 1979), p. 23.

37. Pagden, *European Encounters*, pp. 76–7.

38. N. P. Canny, 'Ideology of English Colonization: from Ireland to America', *William and Mary Quarterly*, ser. 3, vol. 30 (1973), p. 587.

39. Ibid., p. 595.

40. B. Attwood, 'Introduction. The Past as Future: Aborigines, Australia and the (dis)course of History', in Bain Attwood, ed., *In the Age of Mabo: History, Aborigines and Australia* (St Leonards, NSW: Allen & Unwin, 1996), p. ix.

41. Ibid.

42. E. Dark, *The Timeless Land* (London: Collins, 1941), p. 25.

43. A. Lattas, 'Primitivism, Nationalism and Individualism', in B. Attwood and J. Arnold, eds, 'Power, Knowledge and Aborigines', special edition of *Journal of Australian Studies*, vol. 35 (1992), pp. 45–58.

44. Attwood, 'The Past as Future', pp. i–vi.

45. Howard Morphy, 'Empiricism to Metaphysics: In Defence of the Concept of the Dreamtime', in Tim Bonyhady and Tom Griffiths, eds, *Prehistory to Politics: John Mulvaney, the Humanities and the Public Intellectual* (Carlton South, Vic.: Melbourne University Press, 1996), pp. 163–89.

46. Patrick Wolfe, 'On Being Woken Up: the Dreamtime in Anthropology and in Australian Settler Culture', *Comparative Studies in Society and History*, vol. 32 (1991), pp. 197–224.

47. This is a different enquiry from that of Morphy, for example, who writes: 'I will be tracing forward the development of anthropological ideas about Australian Aboriginal religion, in particular the Dreaming, from their beginnings at the turn of the century until the present' (Morphy, 'Empiricism to Metaphysics', pp. 166–7).

48. Cited in Wolfe, 'On Being Woken Up'.

49. Bruce Chatwin, *Songlines* (London: Picador, 1987/88), p. 14.

50. Attwood, 'The Past as Future', p. xii.

51. C. E. Sayers, ed., *Letters from Victorian Pioneers* (Melbourne: Heinemann, 1969; 1st pub. 1898), p. 108.

52. Mrs E. Millett, *An Australian Parsonage, or the Settler and the Savage in Western Australia* (Nedlands, WA: UWA Press, 1980; 1st pub. 1872), p. 145.

53. Ibid., p. 68.

54. *Sydney Gazette*, 1828, p. 3.

55. G. F. James, ed., *A Homestead History, Being the Reminiscences and Letters of Alfred Joyce of Plaistow and Norwood, Port Phillip 1843–1864* (Melbourne: Oxford University Press, 1942), p. 87. Joyce, coming from Norwood, would certainly have known about the famous Norwood gypsies. It is interesting to speculate whether he drew any comparisons between these two different manifestations of the 'nomad' life.

56. Ibid., p. 91.

57. Sayers, *Letters*, p. 380.

58. Street, *The Savage in Literature*, p. 63.

59. *The European Magazine* (London, 1815), p. 374.

60. Mayhew, *London Labour and the London Poor*, vol. 1, p. 2. The term 'apprehension' here denotes understanding rather than fear.

61. G. J. Whitrow, *Time in History: the Evolution of our General Awareness of Time and Temporal Perspective* (Oxford: Oxford University Press, 1988), pp. 8–10. Whitrow's work perpetuates the ethnocentric belief of Victorian writers that Aboriginal children are in some essentialist way impeded from understanding time, as shown in this insulting passage:

'Australian aborigine children of similar mental capacity to white children find it extremely difficult to tell the time by the clock – something that most Western children usually have learned to do successfully by the age of about 6 or 7. The aborigine children can read the hands of the clock as a memory exercise, but they find it difficult to relate the time they read on the clock to the actual time of day' (p. 7). Again, the representation of a certain kind of knowledge as being beyond the childlike capacities of certain peoples bolsters the status of that knowledge and the cultures in which it originates and is 'truly' understood.

62. Brown, *Human Universals*, pp. 27–31.
63. Dark, *The Timeless Land*, pp. 63–8, 161–2.

Chapter 5

1. Morgan Brierley, 'Weather Forecasts and Prophecies', *Notes and Queries*, 11 January 1879, p. 11.
2. Bernard Capp, *Astrology and the Popular Press: English Almanacs 1500–1800* (London: Faber, 1979), p. 246; Marjorie Nicolson, 'English Almanacs and the New Astronomy', *Annals of Science*, vol. 4, no. 1 (15 January 1939), p. 31. A more detailed discussion of the reasons for claiming calendars as a 'woman's genre' can be found in Maureen Perkins, *Visions of the Future: Almanacs, Time, and Cultural Change 1775–1870* (Oxford: Oxford University Press, 1996), pp. 39–45.
3. The increasing number of anniversary celebrations was also part of this process. See Roland Quinault, 'The Cult of the Centenary, c.1784–1914', *Historical Research*, vol. 71, no. 176 (October 1998), pp. 307–23.
4. Joseph A. Kestner, *Mythology and Misogyny: the Social Discourse of Nineteenth-Century British Classical-Subject Painting* (Madison, Wis.: University of Wisconsin Press, 1989), p. 13.
5. Ibid.
6. Ibid., pp. 50–1.
7. Susan P. Casteras, *The Substance or the Shadow: Images of Victorian Womanhood* (New Haven: Yale Center for British Art, 1982), p. 39.
8. Ibid., p. 33.
9. Susan P. Casteras, *Images of Victorian Womanhood in English Art* (Cranbury, NJ: Associated University Presses, 1987), p. 13.
10. Lynda Nead, 'Representation, Sexuality and the Female Nude', *Art History*, Vol. 6 (June 1982), pp. 227–36.
11. John Hadfield, *Every Picture Tells a Story: Images of Victorian Life* (London: The Herbert Press, 1985), p. 71.
12. Casteras, *The Substance or the Shadow*, p. 42.
13. Kestner, *Mythology and Misogyny*, p. 4.
14. Joanna Banham, 'Walter Crane and the Decoration of the Artistic Interior', in Greg Smith and Sarah Hyde, *Walter Crane 1845–1915: Artist, Designer and Socialist* (London: Lund Humphries, 1989), p. 47.
15. Casteras, *Images of Victorian Womanhood*, p. 12.

16. The foremost examples, perhaps, were Linley Sambourne and George du Maurier, but many others drew occasionally for *Punch*. Walter Crane, for example, made one contribution in 1866. See Forrest Reid, *Illustrators of the Eighteen Sixties: an Illustrated Survey of the Work of 58 British Artists* (New York: Dover Publications, 1975; 1st pub. as *Illustrators of the Sixties*, London: Faber & Gwyer, 1928).
17. *The Lady* (1885), quoted in Irene Dancyger, *A World of Women: an Illustrated History of Women's Magazines* (Dublin: Gill & Macmillan, 1978), p. 77.
18. Gleeson White, 'Sandro Botticelli (Filipepi)', *Dome*, vol. 1 (1897), pp. 81–7, cited in Linda Dowling, *The Vulgarization of Art: the Victorians and Aesthetic Democracy* (Charlottesville and London: University Press of Virginia, 1996), p. 116.
19. Ian Small and Josephine Guy, 'The French Revolution and the Avant-Garde', in Ceri Crossley and Ian Small, eds, *The French Revolution and British Culture* (Oxford: Oxford University Press, 1989), pp. 141–55.
20. Jack J. Spector, 'Medusa on the Barricades', *American Imago*, vol. 53, no. 1 (1996), p. 25.
21. Marina Warner, *Monuments and Maidens: the Allegory of the Female Form* (London: Vintage, 1996; 1st pub. Weidenfeld & Nicolson, 1985), pp. 277–8.
22. Marilyn Yalom, *A History of the Breast* (New York: Knopf, 1997), p. 105.
23. Warner, *Monuments and Maidens*, p. 145.
24. Anne McClintock, *Imperial Leather: Race, Gender, and Sexuality in the Colonial Context* (New York: Routledge, 1995), pp. 33, 35 (italics in original).
25. Joseph Conrad, *Lord Jim: a Tale* (Harmondsworth: Penguin, 1957; 1st pub. 1900), p. 294.
26. Ibid.
27. Ibid., p. 292.

Conclusion

1. Donna Haraway, *Primate Visions: Gender, Race, and Nature in the World of Modern Science* (London: Verso, 1992); Mary Poovey, *A History of the Modern Fact: Problems of Knowledge in the Sciences of Wealth and Society* (Chicago and London: University of Chicago Press, 1998); Sandra Harding, *Whose Science, Whose Knowledge: Thinking From Women's Lives* (Milton Keynes: Open University Press, 1991).
2. Poovey, *A History of the Modern Fact*, p. 7.
3. H. K. Bhabha, *The Location of Culture* (London: Routledge, 1994), ch. 4.
4. Samuel Bamford, *The Autobiography of Samuel Bamford*, vol. 1: *Early Days* (ed. W. H. Chalomer; London: Frank Cass, 1967; 1st pub. 1848), p. 258.
5. David Vincent, *Literacy and Popular Culture: England 1750–1914* (Cambridge: Cambridge University Press, 1989), p. 181.
6. Johannes Fabian, *Time and the Other: How Anthropology Makes its Object* (New York: Columbia University Press, 1983), p. 34.
7. My thanks to Gareth Griffiths for this juxtaposition of terms.

Select Bibliography

Primary Sources

Adolphus, John *Observations on the Vagrant Act and Some Other Statutes, and on the Powers and Duties of Justices of the Peace* (London: John Major, 1824).

Bamford, Samuel *The Autobiography of Samuel Bamford*, vol. 1: *Early Days* (ed. W. H. Chaloner; London: Frank Cass, 1967; 1st pub. 1849).

—— *Passages in the Life of a Radical* (London: Macgibbon & Kee, 1967; 1st pub. 1844).

Blackwood's Magazine

Brierley, Morgan 'Weather Forecasts and Prophecies', *Notes and Queries*, 11 January 1879, pp. 10–13.

British Parliamentary Papers

Chambers's Journal

Collins, David *An Account of the English Colony in New South Wales* (2 vols, London, 1802).

Collins, Mortimer *Thoughts in my Garden* (ed. Edmund Yates, 2 vols, London, 1880).

Collins, Wilkie *The Woman in White* (Oxford, 1949; 1st pub. London, 1860).

—— *The Moonstone* (Harmondsworth: Penguin, 1955; 1st pub. 1868).

Conrad, Joseph *The Secret Agent: A Simple Tale* (London: Penguin, 1988; 1st pub. 1907).

Dark, E. *The Timeless Land* (London: Collins, 1941).

—— *Lord Jim: a Tale* (Harmondsworth: Penguin, 1957; 1st pub. 1900).

The Everlasting Circle of Fate or Expounder of Things that are to be; Being a New and Original Mode of Telling Fortunes (Edinburgh, 1840).

The Fortune Teller (1871).

Gentleman's Magazine

The Golden Dreamer, or Dreamer's Interpreter, Clearly Showing how All Things Past, Present and to Come may be Ascertained by means of Dreams (Newcastle upon Tyne, n.d., catalogued by BL as 1850).

Harding, C. G., ed., *The Republican* (London, 1848), in Dorothy Thompson, ed., *Chartism: Working-Class Politics in the Industrial Revolution* (New York: Garland Publishing, 1986).

Howitt, William *Land, Labour, and Gold or Two Years in Victoria* (London: Longman, Brown, Green, Longmans & Roberts, 1855).

James, G. F., ed. *A Homestead History, Being the Reminiscences and Letters of Alfred Joyce of Plaistow and Norwood, Port Phillip 1843–1864* (Melbourne: Oxford University Press, 1942).

Johnson, Mr *The Female Fortune-Teller: a Comedy as it is Acted at the Theatre in Lincoln's-Inn Fields* (London, 1726).

Lang, John Dunmore *Reminiscences of My Life and Times Both in Church and State in Australia for Upwards of Fifty Years* edited by D. W. A. Baker (Melbourne: Heinemann, 1972).

Leland, Charles Godfrey *Gypsy Sorcery and Fortune-Telling* (London, 1891).

McCrae, Hugh, ed. *Georgiana's Journal: Melbourne 1841–1865* (Sydney, 1934, 1966).

Mayhew, Henry *London Labour and the London Poor: a Cyclopaedia of the Condition and Earnings of Those that Will Work, Those that Cannot Work, and Those that Will Not Work* (London: Frank Cass, 1967; 1st pub. 1851).

Millett, Mrs E. *An Australian Parsonage or, the Settler and the Savage in Western Australia* (Nedlands, WA: UWA Press, 1980; 1st pub. 1872).

The New Infallible Fortune Teller, or a just interpretation of Dreams and Moles, to which are added, rules to foretell the Weather; drawn up from the strict observance of nearly half a century (Edinburgh, 1818).

Sayers, C. E., ed. *Letters from Victorian Pioneers* (Melbourne: Heinemann, 1969; 1st pub. 1898).

Seafield, Frank *The Literature and Curiosities of Dreams: A Commonplace Book or Speculations Concerning the Mystery of Dreams and Visions* (2 vols., London, 1865).

Smithson's Northallerton Almanack

The Society for the Suppression of Vice *The Trial of Joseph Powell* (London: 1808).

Stormon, E. J., ed. *The Salvado Memoirs: Historical Memoirs of Australia and Particularly of the Benedictine Mission of New Norcia of the Habits and Customs of the Australian Natives* (Nedlands, WA: UWA Press, 1977).

The True Fortune Teller (Edinburgh, 1850).

Vo-Key's Royal Pocket Index, Key to Universal Time upon the Face of Every Watch and Clock (London, 1901), BL 8560.a.47.

Wells, H. G. 'The So-called Science of Sociology', in *An Englishman Looks at the World* (London: Cassell, 1914).

Z [i.e. Hannah More], *Tawny Rachel, or the Fortune Teller, with Some Account of Dreams, Omens and Conjurers*, Cheap Repository for Religions and Moral Tracts, no. 17 (London, n.d., but catalogued in BL as possibly 1797).

Secondary Sources

Ashcroft, Bill, Gareth Griffiths and Helen Tiffin, eds. *The Post-Colonial Studies Reader* (London: Routledge, 1995).

Ashton, John *A History of English Lotteries* (London: Leadenhall Press, 1893).

Attwood, B. and J. Arnold, eds, 'Power, Knowledge and Aborigines', special edition of *Journal of Australian Studies*, vol. 35 (1992).

Attwood, Bain, ed. *In the Age of Mabo: History, Aborigines and Australia* (St Leonards, NSW: Allen & Unwin, 1992).

Aveni, Anthony F. *Empires of Time: Calendars, Clocks, and Cultures* (New York: Basic Books, 1989).

Bagwell, Philip S. *The Transport Revolution from 1770* (London: Batsford, 1974).

Banham, Joanna 'Walter Crane and the Decoration of the Artistic Interior', in Greg Smith and Sarah Hyde, *Walter Crane 1845–1915: Artist, Designer and Socialist* (London: Lund Humphries, 1989).

Basham, Diana *The Trial of Woman: Feminism and the Occult Science in Victorian Literature and Society* (Basingstoke: Macmillan, 1992).

Baudrillard, Jean 'Hystericizing the Millennium', <http://www.ctheory.com/a-hystericizing_the.html>, originally published as *L'Illusion de la fin: ou La grève des événements* (Paris: Galilée, 1992; trans. Charles Dudas).

Beckford, James A. *Religion and Advanced Industrial Society* (London: Unwin Hyman, 1989).

Bell, Vikki 'Dreaming and Time in Foucault's Philosophy', *Theory, Culture and Society: Explorations in Critical Social Science*, vol. 11, no. 2 (May 1994), pp. 151–64.

Bhabha, H. K. *The Location of Culture* (London: Routledge, 1994).

Bloom, Harold *Omens of Millennium: The Gnosis of Angels, Dreams and Resurrection* (London: Fourth Estate, 1996).

——, ed. *Sigmund Freud's 'The Interpretation of Dreams'* (New York: Chelsea House, 1987).

Bock, Kenneth 'Theories of Progress, Development, Evolution', in Tom Bottomore and Robert Nisbet, eds, *A History of Sociological Analysis* (London: Heinemann, 1979), pp. 39–79.

Bristow, Edward J. *Vice and Vigilance: Purity Movements in Britain since 1700* (Dublin: Gill and Macmillan, 1977).

Brown, Callum G. 'Did Urbanization Secularize Britain?', *Urban History Yearbook* (1988), pp. 1–14.

Brown, Donald E. *Human Universals* (Philadelphia: Temple University Press, 1991).

Burke, Peter *Popular Culture in Early Modern Europe* (London: T. Smith, 1978).

Bury, J. B. *The Idea of Progress: an Inquiry into its Origin and Growth* (London: Macmillan, 1924; 1st pub. 1920).

Bushaway, Bob *By Rite: Custom, Ceremony and Community in England 1700–1880* (London: Junction, 1982).

—— '"Tacit, Unsuspected, but Still Implicit Faith": Alternative Belief in Nineteenth-Century Rural England', in Tim Harris, ed., *Popular Culture in England, c. 1500–1850* (New York: St Martin's Press, 1995), pp. 189–215.

Butler, Marilyn 'Antiquarianism (Popular)', in Iain McCalman, ed., *The Oxford Companion to the Romantic Age: British Culture 1776–1832* (Oxford: Oxford University Press, 1999), pp. 328–38.

Canny, N. P. 'Ideology of English Colonization: from Ireland to America', *William and Mary Quarterly*, ser. 3, vol. 30 (1973).

Casteras, Susan P. *The Substance or the Shadow: Images of Victorian Womanhood* (New Haven: Yale Center for British Art, 1982).

—— *Images of Victorian Womanhood in English Art* (Cranbury, NJ: Associated University Presses, 1987).

Castle, Terry *The Female Thermometer: Eighteenth-Century Culture and the Invention of the Uncanny* (New York: Oxford University Press, 1995).

Chakrabarty, Dipesh 'Minority Histories, Subaltern Pasts', *Perspectives*, vol. 35, no. 8 (November 1997).

Chatwin, Bruce *Songlines* (London: Picador, 1987/88).

Cohen, Stanley 'The Punitive City: Notes on the Dispersal of Social Control', in Stuart Henry, ed., *Social Control: Aspects of Non-state Justice* (Aldershot, Hants: Dartmouth, 1994), pp. 21–46.

Cross, Gary *Time and Money: the Making of Consumer Culture* (London: Routledge, 1993).

Cummins, Pauline *Inis t'Oirr/Aran Dance*, cited in Catherine Nash, 'Reclaiming Vision: Looking at Landscape and the Body', *Gender, Place and Culture* vol. 3, no. 2 (1996), pp. 149–69.

Cunningham, Hugh *Leisure in the Industrial Revolution c.1780–c.1880* (London: Croom Helm, 1980).

Curry, Patrick *Prophecy and Power: Astrology in Early Modern England* (Cambridge: Polity, 1989).

—— 'Astrology in Early Modern England: the Making of a Vulgar Knowledge', in Stephen Pumfrey, Paolo L. Rossi and Maurice Slawinski, eds, *Science, Culture and Popular Belief in Renaissance Europe* (Manchester: Manchester University Press, 1991), pp. 274–91.

—— *A Confusion of Prophets: Victorian and Edwardian Astrology* (London: Collins & Brown, 1992).

Curti, Merle 'The American Exploration of Dreams and Dreamers', *Journal of the History of Ideas*, vol. 27 (1966), pp. 391–416.

Cutt, Margaret Nancy *Ministering Angels: a Study of Nineteenth-Century Evangelical Writing for Children* (Wormley, Herts.: Five Owls Press, 1979).

Dancyger, Irene *A World of Women: an Illustrated History of Women's Magazines* (Dublin: Gill & Macmillan, 1978).

Darwin, John *The Triumphs of Big Ben* (London: Robert Hale, 1986).

Davies, Alun C. 'Greenwich and Standard Time', *History Today*, vol. 28 (1978), pp. 194–9.

Davies, Owen *Witchcraft, Magic and Culture 1736–1951* (Manchester: Manchester University Press, 1999).

—— *A People Bewitched: Witchcraft and Magic in Nineteenth-Century Somerset* (Trowbridge, Wiltshire: Bruton, 1999).

Davison, Graeme *The Unforgiving Minute: How Australia Learned to Tell the Time* (Melbourne: Oxford University Press, 1993).

Dawe, Alan 'Theories of Social Action', in Tom Bottomore and Robert Nisbet, eds, *A History of Sociological Analysis* (London: Heinemann, 1978), pp. 362–417.

Donaldson, Mike 'The End of Time? Aboriginal Temporality and the British Invasion of Australia', *Time and Society* vol. 5, no. 2 (1996), pp. 189–207.

Dowling, Linda *The Vulgarization of Art: the Victorians and Aesthetic Democracy* (Charlottesville and London: University Press of Virginia, 1996).

Dreher, Nan H. 'The Virtuous and the Verminous: Turn-of-the-Century Moral Panics in London's Public Parks', *Albion*, vol. 29, no. 2 (Summer 1997), pp. 246–67.

Eze, Emmanuel Chukwudi *Postcolonial African Philosophy* (Cambridge, Mass.: Blackwell, 1997).

Fabian, Johannes *Time and the Other: How Anthropology Makes its Object* (New York: Columbia University Press, 1983).

Fox, Charles and Marilyn Lake, eds. *Australians at Work: Commentaries and Sources* (Fitzroy, Vic.: McPhee Gribble, 1990).

Freud, Sigmund *The Essentials of Psycho-Analysis* (trans. James Strachey; London: Hogarth Press, 1986).

Galer, Allan M. *Norwood and Dulwich: Past and Present, with Historical and Descriptive Notes* (London: Truslove and Shirley, 1890).

Gandhi, Leela *Postcolonial Theory: A Critical Introduction* (St Leonards, NSW: Allen & Unwin, 1998).

Giddens, Anthony *The Consequences of Modernity* (Cambridge: Polity Press in association with Basil Blackwell, 1990).

—— *Modernity and Self-Identity: Self and Society in the Late Modern Age* (Cambridge: Polity, 1991).

Goleman, Daniel *Emotional Intelligence* (New York: Bantam Books, 1995).

Goodall, Heather *From Invasion to Embassy: Land in Aboriginal Politics in New South Wales, 1770–1972* (St Leonards, NSW: Allen & Unwin, 1996).

Goodwin, Jason *Lords of the Horizons: a History of the Ottoman Empire* (London: Chatto & Windus, 1998).

Greenhouse, Carol J. *A Moment's Notice: Time Politics across Cultures* (Ithaca, NY: Cornell University Press, 1996).

Hacking, Ian 'Nineteenth Century Cracks in the Concept of Determinism', *Journal of the History of Ideas* (1983), pp. 455–475.

Hadfield, John *Every Picture Tells a Story: Images of Victorian Life* (London: The Herbert Press, 1985).

Hall, Catherine and Leonore Davidoff *Family Fortunes: Men And Women of the English Middle Class 1780–1850* (London: Hutchinson, 1987).

Haraway, Donna *Primate Visions: Gender, Race, and Nature in the World of Modern Science* (London: Verso, 1992).

Harding, Sandra 'Is Modern Science an Ethnoscience? Rethinking Epistemological Assumptions', in Emmanuel Chukwudi Eze, *Postcolonial African Philosophy: a Critical Reader* (Cambridge, Mass.: Blackwell, 1997), pp. 45–70.

—— *Whose Science, Whose Knowledge? Thinking From Women's Lives* (Milton Keynes: Open University Press, 1991).

Hewitt, Virginia *Beauty and the Banknote: Images of Women on Paper Money* (London: British Museum Press, 1994).

Hillegas, Mark *The Future as Nightmare: H. G. Wells and the Anti-Utopians* (New York: Oxford University Press, 1967).

Hilton, Tim 'Preface', in Samuel Bamford, *Passages in the Life of a Radical* (London: Macgibbon & Kee, 1967; 1st pub. 1844).

Horn, Pamela *The Rural World 1780–1850: Social Change in the English Countryside* (New York: St Martin's Press, 1980).

Houghton, Walter E. *The Victorian Frame of Mind 1830–1870* (New Haven and London: Yale University Press, 1957).

Howse, Derek *Greenwich Time and the Discovery of the Longitude* (Oxford: Oxford University Press, 1980).

Hutton, Ronald *The Stations of the Sun: a History of the Ritual Year in Britain* (Oxford: Oxford University Press, 1996).

James, Tony *Dream, Creativity, and Madness in Nineteenth-Century France* (Oxford: Clarendon Press, 1995).

Keating, Peter, ed. *Into Unknown England 1866–1913: Selections from the Social Explorers* (London: Fontana, 1976).

Kenrick, Donald and Grattan Puxon *The Destiny of Europe's Gypsies* (London: Heinemann, 1972).

Kern, Stephen *The Culture of Time and Space 1880–1918* (Cambridge, Mass.: Harvard University Press, 1983.

Kestner, Joseph A. *Mythology and Misogyny: the Social Discourse of Nineteenth-Century British Classical-Subject Painting* (Madison, Wisc.: University of Wisconsin Press, 1989).

Koselleck, Reinhart *Futures Past: on the Semantics of Historical Time* (Frankfurt am Main, 1979; trans. Keith Tribe, Cambridge, Mass.: MIT Press, 1985).

Lawrence, D. H. *The Virgin and the Gypsy* (London: Heinemann, 1980).

Lehmann, Hartmut and Guenther Roth, eds. *Weber's Protestant Ethic: Origins, Evidence, Contexts* (New York: Cambridge University Press, 1993).

Letts, Anthony 'A History of Letts', in *Letts Keep a Diary*, catalogue to the 'Exhibition of the History of Diary Keeping in Great Britain from Sixteenth to Twentieth Century in Commemoration of 175 years of Diary Publishing by Letts', The Mall Galleries, London, 28 September–25 October 1987' (London: Charles Letts, 1987).

Leventhal, Herbert *In the Shadow of the Enlightenment: Occultism and Renaissance Science in Eighteenth-Century America* (New York: New York University Press, 1976).

Liégeois, J.-P. *Gypsies, An Illustrated History* (trans. Tony Berrett, London: Al Saqi Books, 1986).

Loeb, Lori Anne *Consuming Angels: Advertising and Victorian Women* (New York: Oxford University Press, 1994).

Lukes, Steven *Individualism* (Oxford: Basil Blackwell, 1973).

McClintock, Anne *Imperial Leather: Race, Gender and Sexuality in the Colonial Contest* (New York: Routledge, 1995).

Macey, Samuel L. *Clocks and the Cosmos: Time in Western Life and Thought* (Hamden, Conn.: Archon Books, 1980).

McGrath, Joseph E. *The Social Psychology of Time: New Perspectives* (Newbury Park: Sage, 1988).

MacKenzie, John M. '"In Touch with the Infinite": the BBC and the Empire, 1923–53', in John M. MacKenzie, ed., *Imperialism and Popular Culture* (Manchester: Manchester University Press, 1986), pp. 165–91.

McKibbin, Ross 'Working-Class Gambling in Britain, 1880–1939', in Ross McKibbin, *The Ideologies of Class: Social Relations in Britain 1880–1950* (Oxford, 1990), pp. 101–39.

Malcolmson, Robert *Popular Recreations in English Society 1700–1850* (Cambridge: Cambridge University Press, 1973).

Martin, Henri-Jean 'The Bibliothèque Bleue: Literature for the Masses in the Ancien Régime', trans. David Gerard, in *Publishing History*, vol. 3 (1978), pp.70–102.

Mayall, David *Gypsy-Travellers in Nineteenth-Century Society* (Cambridge: Cambridge University Press, 1988).

Mendelson, Sara and Patricia Crawford *Women in Early Modern England 1550–1720* (Oxford: Clarendon Press, 1998).

Morphy, Howard 'Empiricism to Metaphysics: In Defence of the Concept of the Dreamtime', in Tim Bonyhady and Tom Griffiths, eds, *Prehistory to Politics: John Mulvaney, the Humanities and the Public Intellectual* (Carlton South, Victoria: Melbourne University Press, 1996), pp. 163–89.

Morrison, Ken *Marx, Durkheim, Weber: Formations of Modern Social Thought* (London: Sage, 1995).

Nash, Catherine 'Reclaiming Vision: Looking at Landscape and the Body', *Gender, Place and Culture*, vol. 3, no. 2 (1996), pp. 149–69.

Nead, Lynda 'Representation, Sexuality and the Female Nude', *Art History*, vol. 6 (June 1982), pp. 227–36.

Nicolson, Marjorie 'English Almanacs and the New Astronomy', *Annals of Science*, vol. 4, no. 1 (15 January 1939), pp. 1–33.

Noland, Aaron 'Proudhon and Rousseau', *Journal of the History of Ideas*, vol. 28, no. 1 (January–March 1967), pp. 33–54.

Obelkevich, James *Religion and Rural Society: South Lindsey 1825–1875* (Oxford: Clarendon Press, 1976).

O'Neil, Mary R. 'Superstition', in Mircea Eliade, ed. in chief, *Encyclopaedia of Religion* (New York: Macmillan, 1987), vol. 14, p. 165.

Oppenheim, Janet *The Other World: Spiritualism and Psychical Research in England 1850–1914* (Cambridge: Cambridge University Press, 1985).

Owen, Alex *The Darkened Room: Women, Power, and Spiritualism in Late Victorian England* (Philadelphia: University of Pennsylvania Press, 1990).

Pagden, Anthony *European Encounters with the New World: from Renaissance to Romanticism* (New Haven: Yale University Press, 1993).

—— *The Fall of Natural Man: the American Indian and the Origins of Comparative Ethnology* (Cambridge: Cambridge University Press, 1982).

Perkins, Maureen *Visions of the Future: Almanacs, Time, and Cultural Change 1775–1870* (Oxford: Oxford University Press, 1996).

Pickering, Paul A. 'Chartism and the "Trade of Agitation" in Early Victorian Britain', *History*, vol. 76 (1991), pp. 221–37.

Pimlott, J. A. R. *The Englishman's Holiday: a Social History* (London: Faber & Faber, 1947).

Poole, Robert *Time's Alteration: Calendar Reform in Early Modern England* (London: UCL Press, 1998).

Poovey, Mary *A History of the Modern Fact: Problems of Knowledge in the Sciences of Wealth and Society* (Chicago and London: University of Chicago Press, 1998).

Price, S. R. F. 'The Future of Dreams: From Freud to Artemidorus', *Past and Present*, no. 113 (November 1986), pp. 3–37.

Purcell, Kate *More in Hope than Anticipation: Fatalism and Fortune Telling amongst Women Factory Workers* (Manchester: University of Manchester Sociology Department, Studies in Sexual Politics, no. 20, 1988).

Quinault, Roland 'The Cult of the Centenary, c.1784–1914', *Historical Research*, vol. 71, no. 176 (October 1998), pp. 307–23.

Quinlan, Maurice J. *Victorian Prelude: a History of English Manners 1700–1830* (London: Frank Cass, 1965; 1st pub. New York: Columbia University Press, 1941).

Reid, Douglas A. 'The Decline of Saint Monday, 1766–1876', *Past and Present*, vol. 71 (May 1976), pp. 76–101.

Reid, Forrest *Illustrators of the Eighteen Sixties: an Illustrated Survey of the Work of 58 British Artists* (New York: Dover Publications, 1975; 1st pub. as *Illustrators of the Sixties*, London: Faber & Gwyer, 1928).

Roberts, M. J. D. 'The Society for the Suppression of Vice and its Early Critics, 1802–1812', *The Historical Journal*, vol. 26, no. 1 (1983), pp. 159–176.
—— 'Public and Private in Early Nineteenth-Century London: the Vagrant Act of 1822 and its Enforcement', *Social History*, vol. 13, no. 3 (1988), pp. 273–94.
Robinson, Eric, ed., *John Clare's Autobiographical Writings* (Oxford: Oxford University Press, 1983).
Rutz, Henry J., ed. *The Politics of Time* (American Ethnological Society Monograph Series, no. 4, Washington DC, 1992).
Sardar, Ziauddin, ed. *Rescuing all our Futures: the Future of Futures Studies* (Westport, Conn.: Praeger, 1999).
Seleski, Patty 'Women, Work and Cultural Change in Eighteenth- and Early Nineteenth-Century London', in Tim Harris, ed., *Popular Culture in England c. 1500–1850* (New York: St Martin's Press, 1995), pp. 143–67.
Small, Ian and Josephine Guy 'The French Revolution and the Avant-Garde', in Ceri Crossley and Ian Small, eds, *The French Revolution and British Culture* (Oxford: Oxford University Press, 1989), pp. 141–55.
Smith, Greg and Sarah Hyde *Walter Crane 1845–1915: Artist, Designer and Socialist* (London: Lund Humphries, 1989).
Smith, Mark *Religion in Industrial Society: Oldham and Saddleworth 1740–1865* (Oxford: Oxford University Press, 1994).
Sobel, Dava *Longitude: the True Story of a Lone Genius who Solved the Greatest Scientific Problem of his Time* (New York: Walker, 1995).
Spadafora, David *The Idea of Progress in Eighteenth-Century Britain* (New Haven, Conn.: Yale University Press, 1990).
Spector, Jack J. 'Medusa on the Barricades, *American Imago*, vol. 53, no. 1 (1996) p. 25.
Spufford, Margaret *Small Books and Pleasant Histories: Popular Fiction and its Readership in Seventeenth-Century England* (London: Methuen, 1981).
Stanner, W. E. H. *White Man Got No Dreaming* (Canberra: Australian National University Press, 1979).
Stiles, Meredith N. *The World's Work and the Calendar* (Boston: The Gorham Press, 1933).
Street, B. V. *The Savage in Literature: Representations of 'Primitive' Society in English Fiction 1858–1920* (London: Routledge & Kegan Paul, 1975).
Thomas, Keith *Religion and the Decline of Magic* (Harmondsworth: Penguin, 1985; 1st pub. London: Weidenfeld & Nicolson, 1971).
Thompson, Dorothy, ed. *Chartism. Working-Class Politics in the Industrial Revolution* (New York, 1986).
Thompson, E. P. 'Time, Work-Discipline and Industrial Capitalism', *Past and Present*, vol. 38 (1967), pp. 56–97.
—— *The Making of the English Working Class* (Harmondsworth: Penguin, 1986; 1st pub. London: Victor Gollancz, 1963).
Urciuoli, Bonnie 'Time, Talk, and Class: New York Puerto Ricans as Temporal and Linguistic Others', in Henry J. Rutz, *The Politics of Time* (Washington DC: American Ethnological Society Monograph Series, No. 4, 1992), pp. 108–126.
Vincent, David *Literacy and Popular Culture: England 1750–1914* (Cambridge: Cambridge University Press, 1989).

—— *Bread, Knowledge and Freedom: A Study Of Nineteenth-Century Working Class Autobiography* (London: Europa, 1981).

Wagar, W. Warren 'Modern Views of the Origins of the Idea of Progress', *Journal of the History of Ideas*, vol. 28 (January–March 1967), pp. 59–69.

Walton, John K. and Robert Poole 'The Lancashire Wakes in the Nineteenth Century', in Robert D. Storch, ed., *Popular Culture and Custom in Nineteenth-Century England* (London: Croom Helm, 1982), pp. 100–24.

Wark, McKenzie 'Spooked by postmodernism', *The Australian*, 18 September 1999, p. 42.

Warner, Marina *Monuments and Maidens: the Allegory of the Female Form* (London: Vintage, 1996; 1st pub. Weidenfeld & Nicolson, 1985).

Weber, Max *Economy and Society*, (Berkeley: University of California Press, 1978; eds Guenther Roth and Claus Wittich).

White, Gleeson 'Sandro Botticelli (Filipepi)', *Dome*, vol. 1 (1897), pp. 81–7, cited in Linda Dowling, *The Vulgarization of Art: the Victorians and Aesthetic Democracy* (Charlottesville and London: University Press of Virginia, 1996).

Whitrow, G. J. *Time in History: the Evolution of our General Awareness of Time and Temporal Perspective* (Oxford: Oxford University Press, 1988).

Wiener, Joel *The War of the Unstampd: the Movement to Repeal the British Newspaper Tax, 1830–1836* (Ithaca, NY: Cornell University Press, 1969).

Wiener, Martin J. *Reconstructing the Criminal: Culture, Law, and Policy in England 1830–1914* (Cambridge: Cambridge University Press, 1990).

Wilson, P. W. *The Romance of the Calendar* (New York: Norton, 1937).

Wolfe, Patrick 'On Being Woken Up: the Dreamtime in Anthropology and in Australian Settler Culture', *Comparative Studies in Society and History*, vol. 32 (1991), pp. 197–224.

—— 'History and Imperialism: A Century of Theory, from Marx to Postcolonialism', *The American Historical Review*, vol. 102, no. 2 (April 1997).

Yalom, Marilyn *A History of the Breast* (New York: Knopf, 1997).

Young, Robert J. C. *Colonial Desire: Hybridity in Theory, Culture and Race* (London: Routledge, 1995).

Index